MATHS & CALCULATORS

Contents

1   **Maths**

49   **Pocket Calculators**

97   **Calculator Puzzles**

144   **Index**

The material in this book is also available
in three separate books with titles:
Introduction to Maths, The Pocket Calculator Book
and Practise Your Calculator Skills.

INTRODUCTION TO
MATHS

Nigel Langdon and Janet Cook

Designed by Kim Blundell and Sue Mims

Illustrated by Jane Andrews, Jeremy Banks, Martin Newton
Simon Roulstone, Chris Lyon, Iain Ashman
and Naomi Reed

Computer programs by Chris Oxlade

Edited by Lisa Watts

WITH COMPUTER PROGRAM LISTINGS

Contents

- 3 What is maths?
- 4 The world of numbers
- 6 Decimals
- 8 Zero and negatives
- 10 Use your head
- 12 Shapes everywhere
- 14 Measuring surfaces
- 16 Circles
- 18 3-D
- 20 Angles
- 22 Topology
- 24 Fractions
- 26 Ratios
- 28 Sets
- 30 Statistics
- 32 Graphs
- 34 More about graphs
- 36 Geometry
- 38 Number patterns
- 40 Binary numbers
- 42 Probability
- 45 Computer programs
- 46 Puzzle answers and score chart
- 48 Maths words

First published in 1985 by Usborne Publishing Ltd, 20 Garrick Street, London WC2E 9BJ, England.
Copyright © 1985 Usborne Publishing
The name Usborne and the device are Trade Marks of Usborne Publishing Ltd. All rights reserved. No part of this publication may be reproduced, stored in a retrieval system or transmitted in any form or by any means, electronic, mechanical, photocopying, recording or otherwise, without the prior permission of the publisher.

What is maths?

The word mathematics comes from the Greek word *manthanein* meaning to learn. This is a clue to what maths is all about. For more than 4000 years, from the building of the pyramids to the designing of a modern motorway intersection, people have been studying the relationships between shapes and numbers to help them to learn about and organize the world we live in.

Orbiting satellites in space, repeating patterns in the design of wallpaper, driving a car – all depend upon the mathematics of shape, number and precision. What connects these different features is logic, and studying maths can help you to think more clearly and logically.

The best way to understand maths is to take a pencil and paper (or a calculator) and try things out for yourself. There are a number of puzzles scattered throughout this book which you can do to test yourself. The answers are at the back of each section.

If you are not familiar with them, mathematical symbols may look a bit frightening. In fact, they are just a shorthand way of writing simple statements. Once you have learned the basics, you will find it easier to understand the more complicated ideas. So don't expect to understand everything at once; tackle small sections at a time.

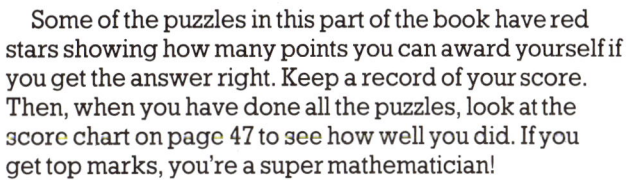

Some of the puzzles in this part of the book have red stars showing how many points you can award yourself if you get the answer right. Keep a record of your score. Then, when you have done all the puzzles, look at the score chart on page 47 to see how well you did. If you get top marks, you're a super mathematician!

On page 45, there are some simple computer programs to test mathematical theories such as probability. The programs will work on most types of home computer. If you have, or can borrow, a computer, try them out – they are very short and take only a few minutes to type in.

The world of numbers

Can you imagine a world without numbers? Numbers are such an important part of everyday conversation that it would be extremely difficult to go shopping or tell the time without using them. Think of any recent developments in modern technology, for example satellites, lasers or cable television; they would have been impossible without the science of numbers.

How many?

The answer to the question "How many?" is always a number. Yet numbers on their own can be very misleading.

How far is Austria from France?

5
1
200
125

The answers are all correct. Can you think of words which could accompany each of these numbers?*

Which one?

Numbers are also used for labelling and identification. Buses are labelled so that all those on the same route have the same number. But each vehicle has a different registration number so that it can be identified.

Counting in tens...

There have been numerous ways of counting in the past, yet almost all of them are based upon counting in tens. This is possibly because most people count on their fingers at some time.

Aborigines of South Australia use a word for 10 which also means "two hands".

...and hundreds and thousands

Most of the ancient number systems had different symbols for units, tens, hundreds and thousands. Can you translate this Egyptian number?

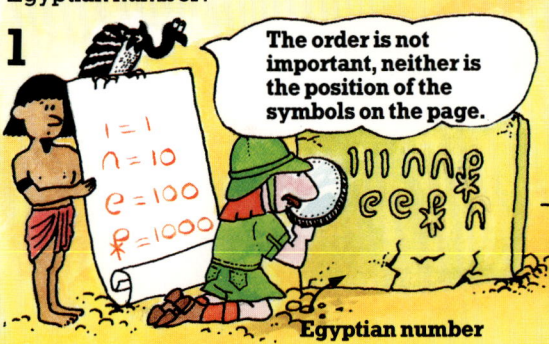

1

I = 1
∩ = 10
℮ = 100
₽ = 1000

The order is not important, neither is the position of the symbols on the page.

Egyptian number

3

C and M are the first letters of centum and mille, the Latin words for hundred and thousand.

ROMA
MCXX

Later, Romans used V for 5 and L for 50.

It was not until the Middle Ages that Latin scholars used IV to mean 4 (one less than 5). How would you translate CM?

Puzzle answers are on pages 46-47.

In Northern Japan, the ancient Ainu language uses *wan* (meaning "two-sided") for 10.

The symbols we use to write numbers are called digits, coming from the Latin for finger.

2 The Romans used I, X, C and M. The symbols I and X (1 and 10) probably came from a much earlier method of writing numbers when cuts were made on a wooden stick.

What is the value of this number?

4 One of the 2s on this sign means 2000 and the other is only 2. Which is which?

India 2192

Our modern decimal system, developed in India by AD 570, is the most sophisticated of all. It is possible to write any number using only ten digits. This is because of the importance given to the order of writing the digits; for example 27 is a different number from 72.*

Pictures by numbers

Just as a colouring book can tell you which colours to use, the pictures sent back to Earth by space ships and satellites are transmitted by numbers.

A Mariner spacecraft which is photographing a planet takes television pictures. Each tiny square of the picture is examined by an on-board computer and given a number between 0 and 63 to indicate its brightness. These numbers are signalled back to Earth and another computer then uses them to reconstruct the picture.

Digital noughts and crosses

A good variation on the game noughts and crosses is to use the digits 0 to 9.

Each digit may be used only once in a game. The winner is the first person to make a line of three numbers which total 10.

Is there a biggest number?

Imagine a very, very large number — the largest you can think of. Can you add 1 to it?

Mathematicians use the symbol ∞ (infinity) to represent quantities which could never be counted, even if you counted for ever.

A dead 8!

*Go to pages 6-7 for more about the decimal system.

Decimals

The decimal system allows us to write any whole number using a combination of up to ten different symbols: 0, 1, 2, 3, 4, 5, 6, 7, 8 and 9. The value of each symbol changes according to its position in the number.

This seven means seven hundreds.

This seven means seven tens.

```
1000s  100s  10s  1s
         9    9   9
                  1 +
   1     0    0   0
```

Reading from the right, the first digit shows the number of 1s, the second digit the number of 10s and so on. You can count up to 9 before moving to the next position.

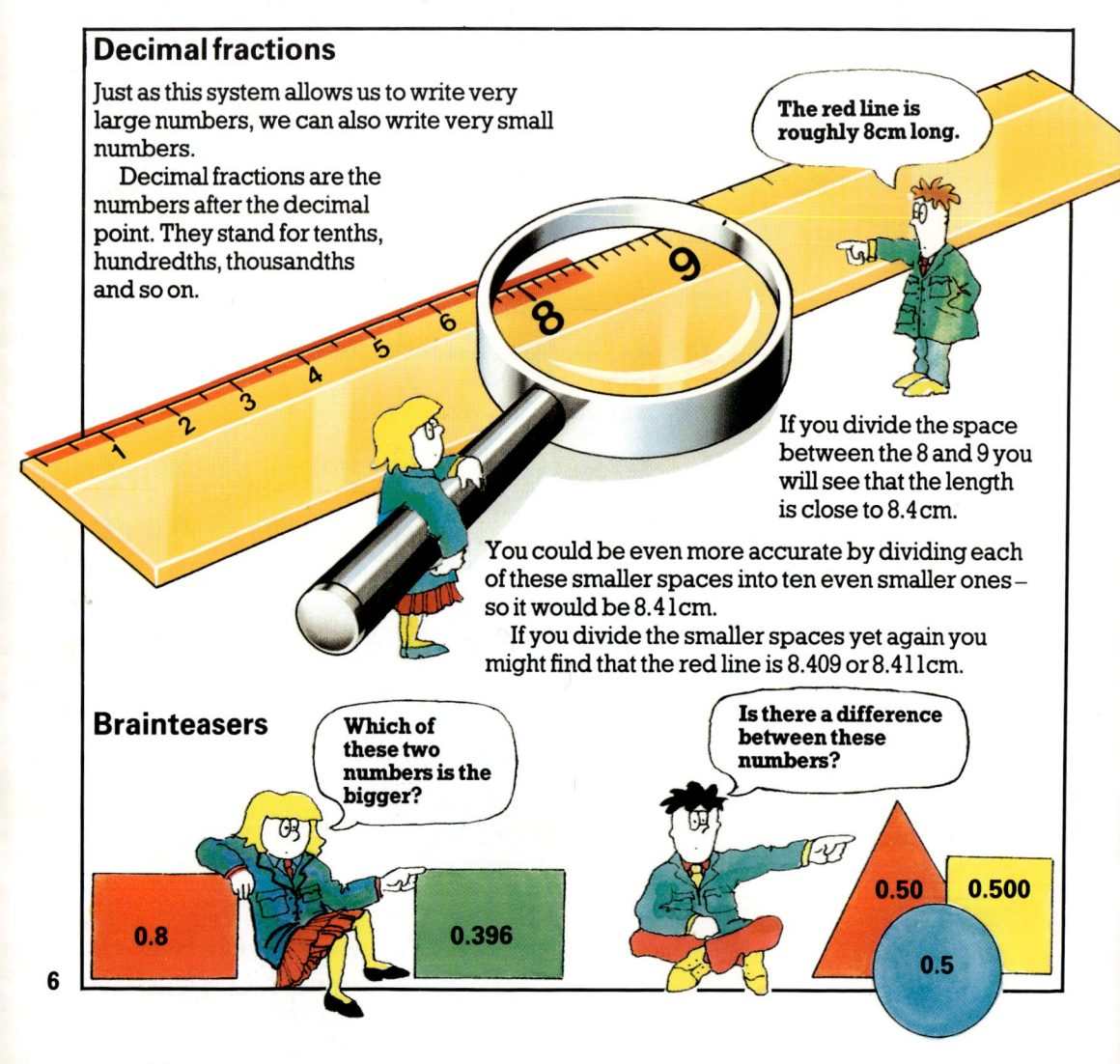

Decimal fractions

Just as this system allows us to write very large numbers, we can also write very small numbers.

Decimal fractions are the numbers after the decimal point. They stand for tenths, hundredths, thousandths and so on.

The red line is roughly 8cm long.

If you divide the space between the 8 and 9 you will see that the length is close to 8.4cm.

You could be even more accurate by dividing each of these smaller spaces into ten even smaller ones – so it would be 8.41cm.

If you divide the smaller spaces yet again you might find that the red line is 8.409 or 8.411cm.

Brainteasers

Which of these two numbers is the bigger?

0.8   0.396

Is there a difference between these numbers?

0.50   0.500   0.5

Rounding off

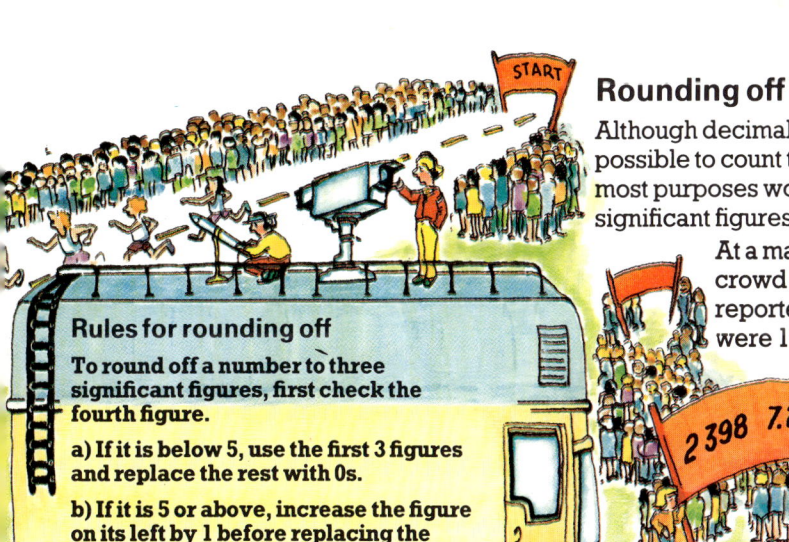

Although decimal fractions make it possible to count to the minutest detail, for most purposes working with three significant figures is accurate enough.

At a marathon race where the crowd totalled 156 432, a reporter might write that there were 156 000 spectators.

2 398   7.2153

Can you round off these numbers?

Rules for rounding off

To round off a number to three significant figures, first check the fourth figure.

a) If it is below 5, use the first 3 figures and replace the rest with 0s.

b) If it is 5 or above, increase the figure on its left by 1 before replacing the rest with 0s.

Moving the decimal point

The two numbers below are roughly 300 and 3, so the first number is 100 times larger than the second.

297.4 ×100 / ÷100 2.974

When you multiply 2.974 by 100 it looks as though the decimal point jumps two places to the right.

In fact, is it the decimal point or the number which moves?

DECIMAL POINT MOVES:
- ÷10 — 1 place left
- ÷100 — 2 places left
- ÷1000 — 3 places left
- ×10 — 1 place right
- ×100 — 2 places right
- ×1000 — 3 places right

Dividing by decimals

When dividing by decimals without a calculator it is easier to move the decimal points so that both the numbers you are working with are whole numbers.

35 ÷ 0.7 is the same as **350 ÷ 7.**

When you are multiplying and dividing on a calculator it is easy to press a wrong key. Use this guide to check that your answer is roughly what you expected.

Dice game

This is a game for two players who take it in turns to throw a dice.

Multiply any number from this table by your dice score.

2.5	3	0.05
1.5	2	0.75
0.5	1	0.25

See if there is a number which matches your answer on the board on the right and cover it.

The first player to cover three numbers in a line is the winner.

3	15	12	3
0.25	2.5	3	0.75	1.5	5
6	2	2.5	1	6	0.05
1.5	0.5	1.25	1	2	0.75
1	3	0.25	0.5	3	1
0.2	1.5	6	4.5	3	4
1.5	12	2	7.5	1.5	

7

Zero and negatives

What is zero? Is it nothing or is it something? Can you have less than zero? The answers may seem obvious to us now, but they puzzled early mathematicians. Because of this, the use of a zero symbol originated quite late in mathematics and it was not until the Middle Ages that it was introduced into Europe from India.

These zeros mean no 100s, no 10s and no 1s, and they make the digit 1 stand for 1000.

Zero puzzled early mathematicians because on its own, it stands for nothing. Yet it has the power to change other numbers when placed next to them.

Zero can also stand for a fixed level. For example, 0°C is the temperature at which water freezes.

Negative numbers are used to show amounts which are below a fixed level. For example, ⁻11°C indicates a temperature 11 degrees lower than freezing point.

The rocket booster has separated at lift-off minus 30.

Times are often given as positive or negative numbers with reference to an important moment, such as a rocket launching into space.

Sums with negative numbers

Sums with negative numbers can be confusing and you may find the number path below helpful.

When doing any sum, you should always look for two things: the instructions and the labels.

⁻4 + 4
Label Instruction

There is always an instruction in a sum but if a number has no label, it is positive.

4 + 4

Using the number path

To do the sum ⁻3−5, start on the stripe for ⁻3. As you are subtracting a number, face left. The sum tells you to subtract 5, and as 5 is positive, move forward five places.

The square you arrive at is ⁻8, so that is the answer.

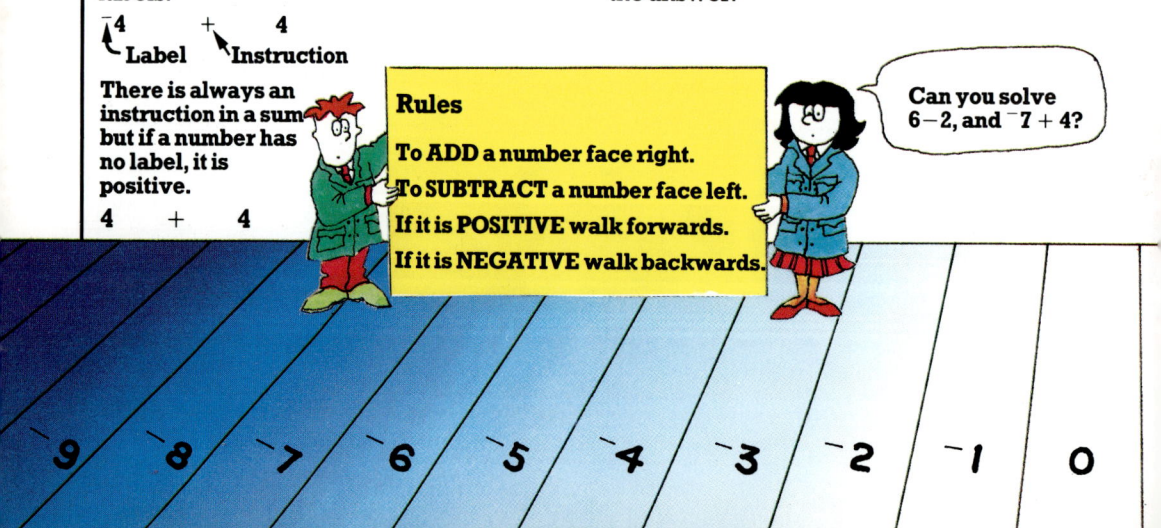

Rules

To ADD a number face right.
To SUBTRACT a number face left.
If it is POSITIVE walk forwards.
If it is NEGATIVE walk backwards.

Can you solve 6−2, and ⁻7 + 4?

⁻9 ⁻8 ⁻7 ⁻6 ⁻5 ⁻4 ⁻3 ⁻2 ⁻1 0

Vectors

One of the important uses of negative numbers is in work with vectors. Vectors are quantities which are represented by arrows to show their direction and size.

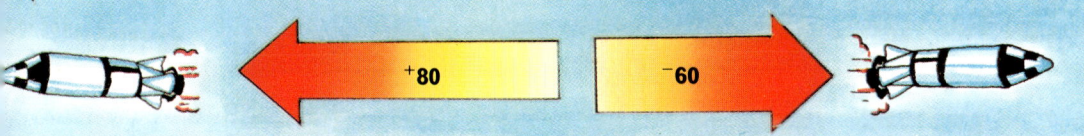

⁺80

⁻60

A rocket which is travelling at 80km per second towards the west could be represented like this. The length of the arrow shows the size of velocity (speed).

Another rocket, travelling at 60km per second due east would be shown like this. The negative symbol means it is travelling in the opposite direction to the first rocket.

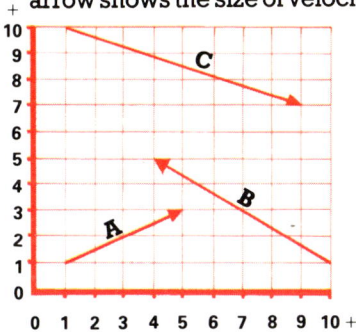

On a grid system like this, vectors are referred to by two numbers. Vector A is ($\binom{4}{2}$) because it goes 4 squares right and 2 squares up. What are vectors B and C?

Vector puzzle

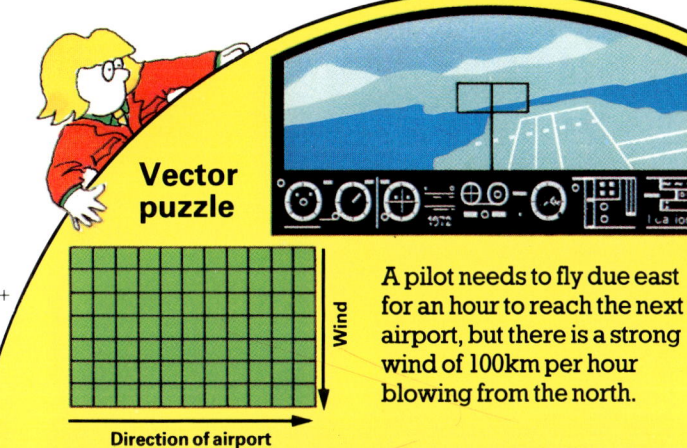

A pilot needs to fly due east for an hour to reach the next airport, but there is a strong wind of 100km per hour blowing from the north.

If he steers towards the airport, the wind will blow him too far south. Can you use this vector diagram to find out the direction in which the pilot should steer?

Multiplication

You can also use the path to do multiplication because when you multiply you are in fact adding up (or if the first number is negative, subtracting) several times. The only difference is that you should start on 0.

$3 \times 5 = (0) + 5 + 5 + 5 = 15$
$3 \times {}^-5 = (0) + {}^-5 + {}^-5 + {}^-5 = {}^-15$
${}^-3 \times {}^-5 = (0) - {}^-5 - {}^-5 - {}^-5 = 15$

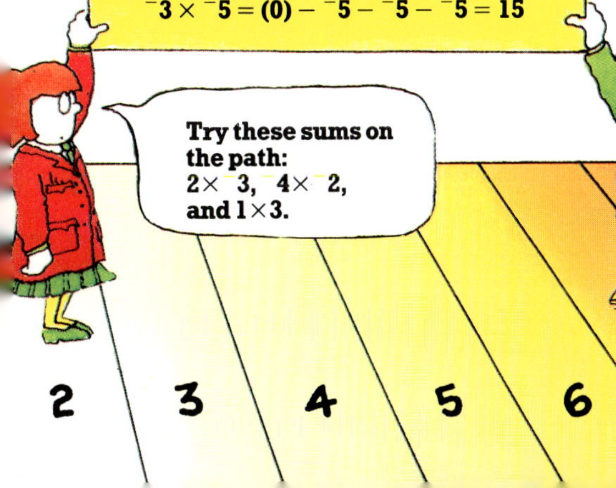

Try these sums on the path:
$2 \times {}^-3$, $4 \times {}^-2$, and 1×3.

Quick tips

If you check your answers, you will find that the following rules always apply:

⁺ × ⁻ = ⁻
⁻ × ⁺ = ⁻
⁺ × ⁺ = ⁺
⁻ × ⁻ = ⁺

Use your head

You don't need to be a genius to be good at mathematics. What you do need is a familiarity with numbers and shapes. This helps you to recognize patterns amongst them and is much better than trying to remember lots of rules by heart. On these two pages there are tips to help you multiply and divide.

There is no set way to calculate with numbers; you should decide for yourself which method allows you to find the correct answer most easily. Can you think of more short cuts yourself?

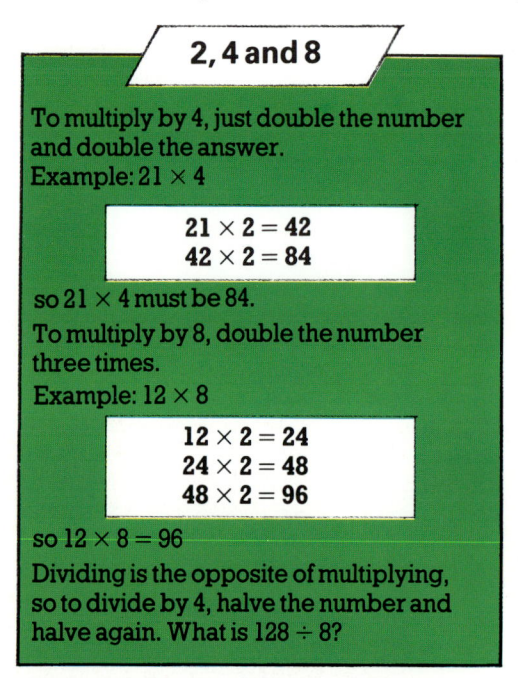

2, 4 and 8

To multiply by 4, just double the number and double the answer.
Example: 21 × 4

> 21 × 2 = 42
> 42 × 2 = 84

so 21 × 4 must be 84.

To multiply by 8, double the number three times.
Example: 12 × 8

> 12 × 2 = 24
> 24 × 2 = 48
> 48 × 2 = 96

so 12 × 8 = 96

Dividing is the opposite of multiplying, so to divide by 4, halve the number and halve again. What is 128 ÷ 8?

Fairground number cruncher

Someone who was very clever at working out simple methods to solve complicated problems was George Bidder, who became a famous fairground entertainer. In 1815, at the age of nine, he took less than a minute to answer this question:

The moon is 123 256 miles from Earth and sound travels at the speed of 4 miles a minute.* How long would it be before someone on the moon could hear an explosion on Earth?

21 days 9 hours and 34 minutes

Multiply by 9

Look at the nine times table:

> 1 × 9 = 9
> 2 × 9 = 18
> 3 × 9 = 27
> 4 × 9 = 36
> 5 × 9 = 45
> 6 × 9 = 54
> 7 × 9 = 63
> 8 × 9 = 72
> 9 × 9 = 81
> 10 × 9 = 90

1 + 8 = 9

2 + 7 = 9

The two digits in the answers always add up to 9. Does this happen when you multiply large numbers by 9?

> 53 × 9 = 477

> 4 + 7 + 7 = 18
> and 1 + 8 = 9

It takes two stages but the digits still add up to 9. This is a useful check and is also helpful if you want to divide by 9. Is 684 divisible by 9?

Multiply by 11

The 11 times table is probably the easiest to recognize:

> 22, 33, 44, 55, 66, 77, 88, 99 . . .

Here is a quick way to multiply 11 by a two-figure number:

> 27 × 11 = 297

Add the 2 and the 7.

Put the answer in between.

Try it with some other numbers.

*We now know that the speed of sound is about 12.5 miles per minute.

Computer brain

In today's world of calculators and computers the human brain can still work amazingly quickly. Shakuntala Devi's is probably the fastest in the world. She lives in Bangalore in India, but spends much of her time travelling around the world giving demonstrations of her ability.

In one famous lecture in Texas, she was asked to find the number which, when multiplied by itself 23 times, gave this answer:

9167486769200391580986609275853801624831066801443086 2240712651642793465704086709659327920576748080679002 2783016354924852380335745316935111903596577547340075 6816883056208210161291328455468057801588067771

Shakuntala worked it out in fifty seconds. In order to check her answer, students used a Univac 1108 computer. It took the computer one full minute to confirm she was right – but it had to be given more than 13 000 instructions first.

(speech bubble: 546372891)

Multiply by 10

To multiply by 10, just add a nought to change the position of the digits.

Example: 25 × 10

```
25 × 10
250
```

To multiply any number by 100, just add two noughts...

```
32 × 100
3200
```

...because multiplying by 100 is the same as multiplying by 10 twice. How would you multiply by 1000?

Multiply by 15

To multiply by 15, you need to recognize that 15 is 10 + half of 10.

Example: 32 × 15

32 × 10	=	320
32 × 5	=	160

Add the two answers: **480**

So 32 × 15 must equal 480.

Try working out these sums:
a) 8 × 15
b) 36 × 15
c) 92 × 15

Beat the teacher

When the teacher of Karl Gauss wanted to keep his class quiet for an hour, he set them questions like the one opposite.

Most of Karl's classmates worked them out slowly on their slates, but Karl, who was only nine, came up with the answers in seconds.

(speech bubble: Work out the sum of 1 + 2 + 3 ... up to 1000.)

Karl, born in 1777, was to become one of the greatest mathematicians ever. He used a neat trick to solve this sum. Can you see what it was? And find the answer?

(speech bubble: Psst: 1 + 1000, 2 + 999, 3 + 998, etc. all equal 1001.)

Shapes everywhere

Mathematics is often called the science of numbers, but that is less than half of it. It is just as much to do with studying shapes and classifying them into particular groups. Once you have done this you can find rules which will apply to solving problems for any shape in the group.

Everything in the world has a shape, from the most intricate snowflake to a skyscraper. The easiest shapes to classify have straight lines, and you can find out about them on these two pages.

**How tall?
How many leaves?
Leaf shape?**

The organization of things into groups is important in many other subjects, for example in the identification of a tree species.

Classifying shapes

Shapes are named according to how many sides and angles they have. The shapes shown here have three or more angles and are called polygons which means "many angles" in Greek. A polygon is a closed, flat shape drawn with straight sides. Using the list below, can you name these polygons?

Triangle means three-angled (Latin)
Quadrilateral means four-sided (Latin)
Pentagon means five-angled (Greek)
Hexagon means six-angled (Greek)

Go to pages 18-19 for a study of three - dimensional objects and more complicated shapes.

Regular polygons

This is the headquarters of the American defence forces.

A regular polygon is one in which all the angles and sides are the same.

How many sides would a polygon need before it became a circle?

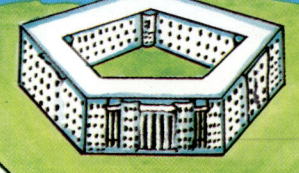

Do you know what it is called?

The family of triangles

Triangles can be grouped according to their sides.

A scalene triangle has all sides different in length.

An isosceles triangle has two sides the same in length.

An equilateral triangle has all sides equal in length.

Can you draw a triangle whose sides are 5cm, 3cm and 9cm?

Squares and rectangles

It is not so easy to organize four-sided polygons into groups.

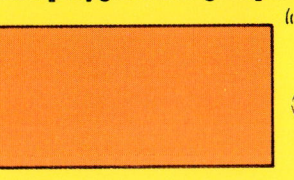

A rectangle is a quadrilateral which has square corners and whose opposite sides are parallel . . .

. . . but that makes a square a rectangle too.

A parallelogram has opposite sides parallel as well, so that makes a square a parallelogram too.

The same shape

◀ Are these two shapes the same? If you picked up the one on the right, turned it over and placed it on top of the shape on the left, you would find that they fitted together exactly. This means that they are congruent because they have identical sides and angles.

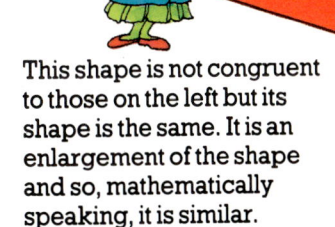

This shape is not congruent to those on the left but its shape is the same. It is an enlargement of the shape and so, mathematically speaking, it is similar.

Fitting shapes

You can make very attractive patterns (called tessellations) by placing different shapes side by side. The Romans used to decorate their walls and floors with "tessellae", which were tiny pieces of mosaic.

Which three regular polygons were used to create the tessellation above?

Measuring surfaces

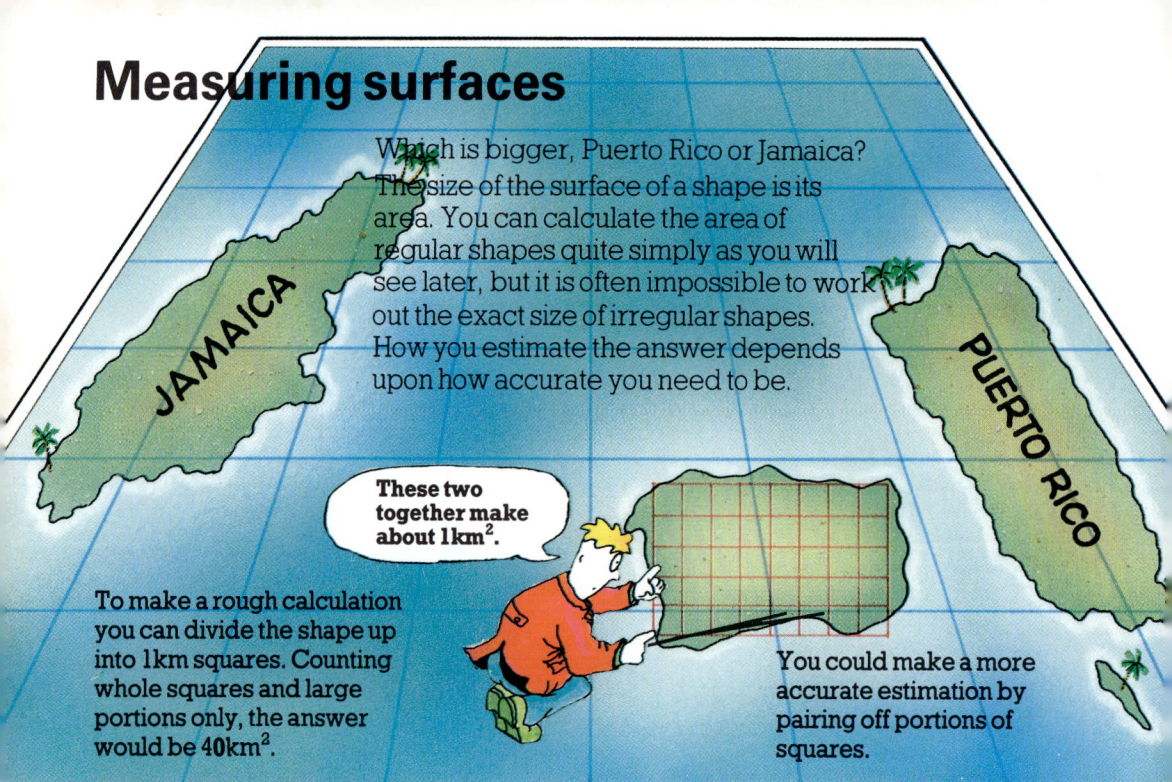

Which is bigger, Puerto Rico or Jamaica? The size of the surface of a shape is its area. You can calculate the area of regular shapes quite simply as you will see later, but it is often impossible to work out the exact size of irregular shapes. How you estimate the answer depends upon how accurate you need to be.

These two together make about 1km².

To make a rough calculation you can divide the shape up into 1km squares. Counting whole squares and large portions only, the answer would be 40km².

You could make a more accurate estimation by pairing off portions of squares.

Rectangular areas

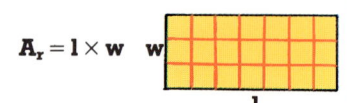

$A_r = l \times w$

If the shape is a rectangle you can count up the squares more quickly by multiplying the length by the width. The area of this rectangle is $7 \times 3 = 21$ squares.

So the formula to calculate the area of a rectangle is $A_r = l \times w$. What do the letters represent?

Rearranging crosses

It is often easier to find the area of a complicated shape by cutting it up and rearranging it to form a more regular outline. The shape still has the same area, as it has just been put together in a different way.

▶If you cut this cross along the red lines you can rearrange it to form the square on the right. You can then use the rectangle formula to find its area.

◀Trace these two crosses and cut around the outlines and along the red lines. Now fit the pieces together. You will find that they also form squares.

Units of measurement

Centimetre squares (cm²) or square inches are used to measure the area of small items like floor tiles.

Square metres (m²) or square yards are the best size for measuring pieces of material.

Area formulae

Because rearranging a shape does not change its area, you can use the rectangle formula when calculating the areas of other shapes.

1. Parallelograms

If you cut off one end of a parallelogram on a line at right angles to its sides and stick this at the other end, you will make a rectangle.

$A_p = l \times b$

The p in the formula stands for parallelogram.

This line is perpendicular to the base because it is at right angles to it.

The area of a parallelogram is the same as the area of a rectangle which has the same side length (l) and the same perpendicular width (b).

2. Triangles

You can draw a rectangle around any triangle using one of the sides as the base of the rectangle.

The base is called b. The height of the rectangle is the same as the height of the triangle from that base.

$A_t = \frac{1}{2}(b \times h)$

Trace the shapes and cut them out to check that they make two equal triangles.

The area of the triangle is half the area of the rectangle.

3. Quadrilaterals

Any quadrilateral can be cut into two triangles, so you can calculate its area in two parts.

A_q = Area of triangle A plus Area of triangle B

Puzzle

How would you find the area of this plan of a sports centre?

Hectares (squares 100m by 100m) or acres (4840 sq. yds) are used for measuring the area of farm land.

Kilometre squares (km^2) or square miles would be used by a geographer measuring the area of an island.

Circles

Ever tried drinking out of a square mug...

... or cycling on hexagonal wheels...

... or using a telephone with a triangular dial?

A circle is a very special shape; it has no corners and any point on its outline (called its circumference) is always the same distance from its central point. This is of great practical use as you can see. It also makes it possible to find a formula to calculate a circle's circumference and area.

Measuring the circle

People have known for centuries that when a wheel turned a complete circle, it always moved forward just over three times its diameter.

$C = \pi d$ or $2\pi r$

In fact, it is about 3.14 times. The number cannot be written exactly, but because of its importance it has been given a name of its own: the Greek letter π (pronounced pie). So to find a circle's circumference, multiply its diameter (or twice the radius) by 3.14.*

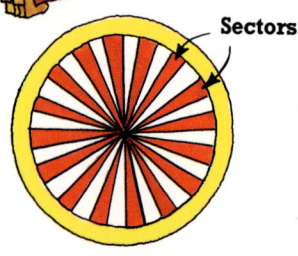

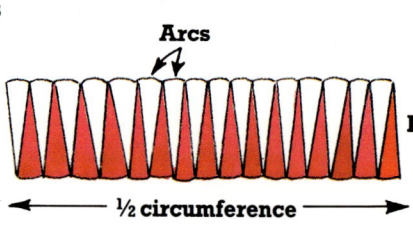

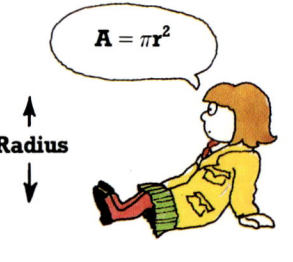

Sectors

Arcs

Radius

½ circumference

$A = \pi r^2$

If you slice a circle into small sectors you can rearrange the circle approximately as a rectangle. Half the arcs are at the top and half are at the bottom.

The width of the rectangle is approximately the radius of the circle, and the length of the rectangle is almost half the circumference. So the area of the circle is $\pi r \times r$ or πr^2.

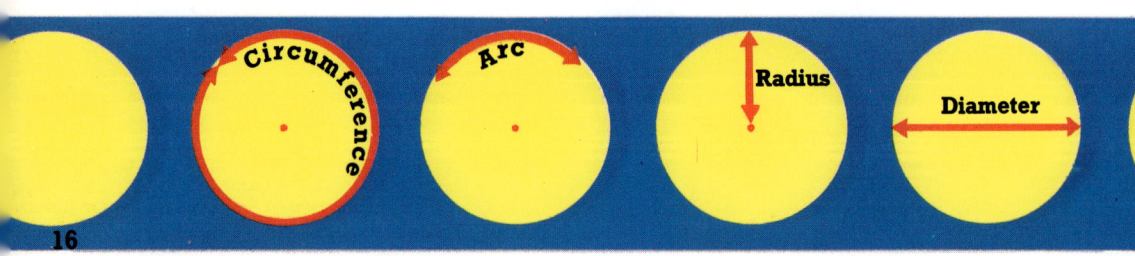

Circumference | Arc | Radius | Diameter

*For more about π, see page 122.

Circular shapes

The cone, cylinder and sphere all have curved surfaces. If you cut straight across any one of them on a line parallel to its base, you would find a circle.

A cylinder

A sphere

A cone

Circle – any slice parallel to the base.

Ellipse – any slice between the circle and parabola.

Parabola – if the cone is sliced parallel to the sloping edge, the cross-section is always a parabola.

Hyperbola – any slice cut at an angle greater than the slope.

How would you cut a cone to produce a triangle?

Hypatia was a famous Greek mathematician who lived in Alexandria around AD 400. She spent many hours examining the curves which could be obtained by cutting through a cone.

She found that a cone could be cut to make three other curves apart from the circle, and that like the circle, they all have their own formulae.

Drawing a circle with straight lines

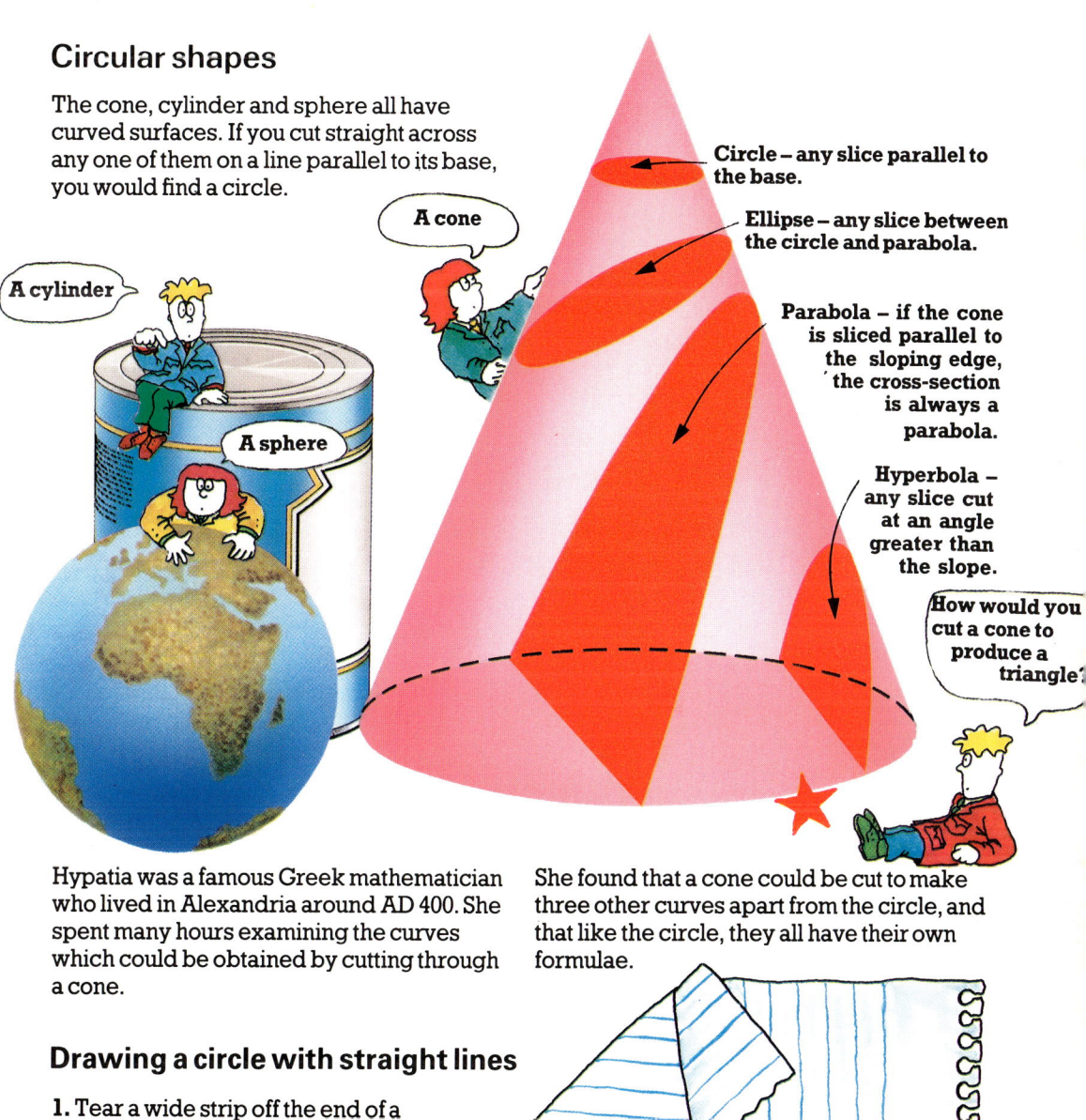

1. Tear a wide strip off the end of a piece of paper and place a cross in roughly the middle of the large piece.

2. Place the strip so that it just touches the cross and fold the edge of the large piece over it.

3. Repeat this about 20 times, placing the strip at a different angle each time.

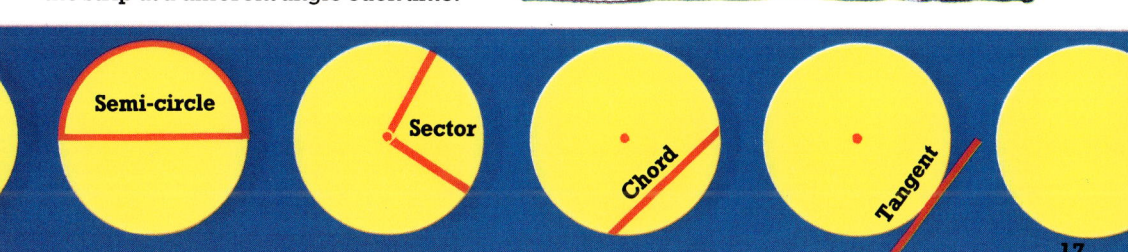

Semi-circle Sector Chord Tangent

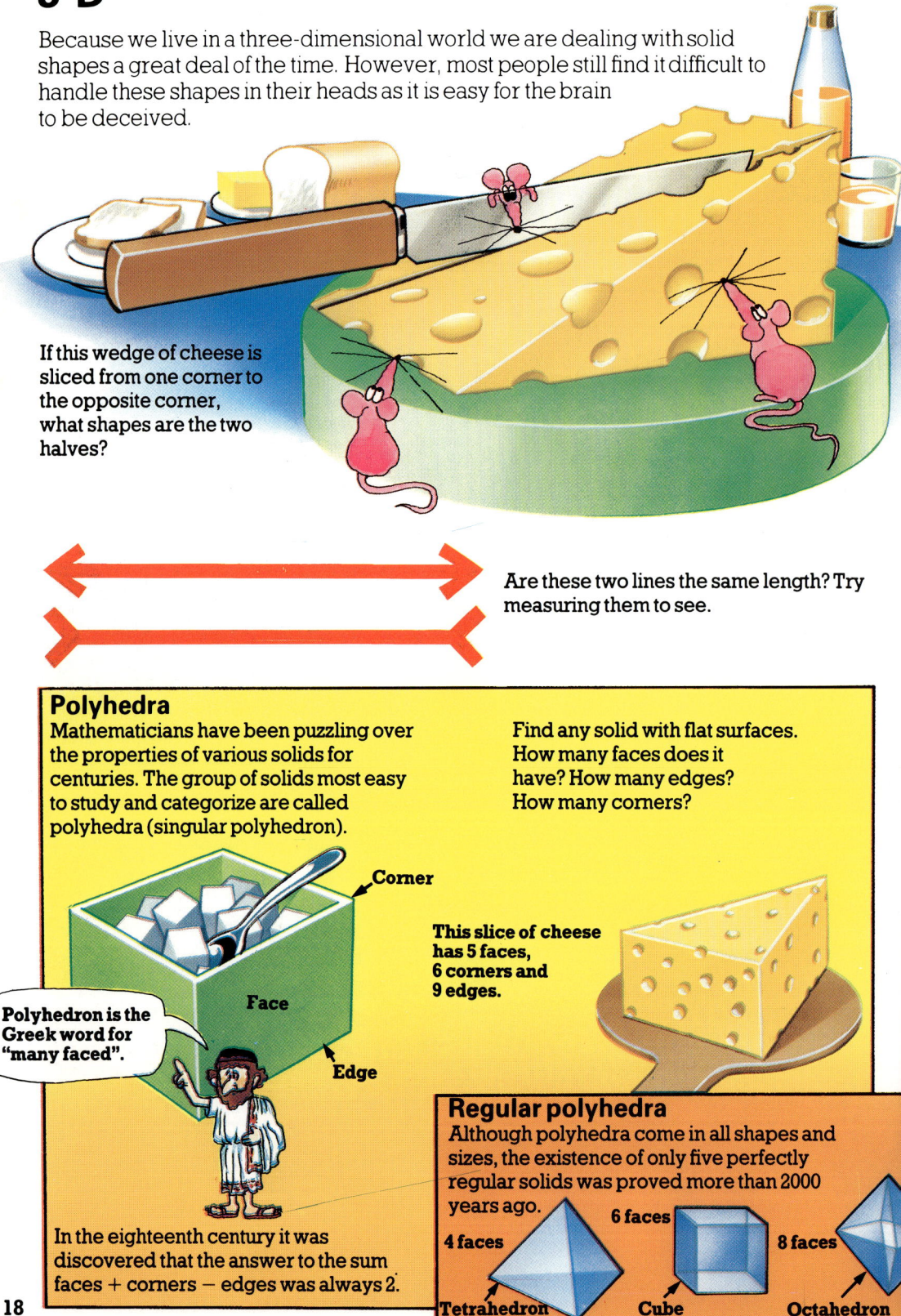

3-D

Because we live in a three-dimensional world we are dealing with solid shapes a great deal of the time. However, most people still find it difficult to handle these shapes in their heads as it is easy for the brain to be deceived.

If this wedge of cheese is sliced from one corner to the opposite corner, what shapes are the two halves?

Are these two lines the same length? Try measuring them to see.

Polyhedra

Mathematicians have been puzzling over the properties of various solids for centuries. The group of solids most easy to study and categorize are called polyhedra (singular polyhedron).

Find any solid with flat surfaces. How many faces does it have? How many edges? How many corners?

Corner

Face

Edge

Polyhedron is the Greek word for "many faced".

This slice of cheese has 5 faces, 6 corners and 9 edges.

In the eighteenth century it was discovered that the answer to the sum faces + corners − edges was always 2.

Regular polyhedra

Although polyhedra come in all shapes and sizes, the existence of only five perfectly regular solids was proved more than 2000 years ago.

4 faces — Tetrahedron
6 faces — Cube
8 faces — Octahedron

Calculating volume

Just as it is difficult to visualize three-dimensional objects, it is hard to estimate how much space an object contains. Mathematics can help you to calculate the space accurately.

The amount of space an object occupies is its volume. For small solids, the volume is measured in cubic centimetres or cubic inches. One cubic centimetre is 1cm × 1cm × 1cm, or 1cm³.

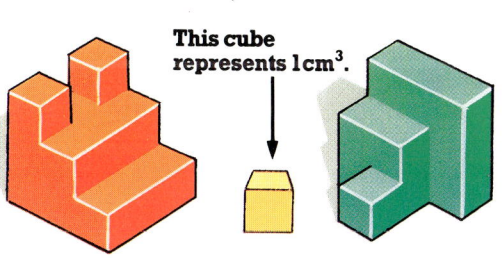

Which suitcase is bigger?

This cube represents 1cm³.

How many cm³ do these two shapes contain?

This fishtank is a cuboid (a solid rectangle). You can work out its volume using the formula v = l × w × h (volume = length × width × height).

You would need 18 cubes to fill one layer of the tank which is 3 cubes wide and 6 cubes long.

You would need four layers to fill the fish tank so its volume is 4 × 18 cubes = 72 cubes.

Another way of putting this is that the volume is the area of the shape's base multiplied by its height. You can use this method to find out the volume of any shape which has the same cross-section throughout.

Can you work out a formula to find the volume of a cylinder?*

Building cubes

Six pairs of these polyhedra will build six cubes. Which one is left over?

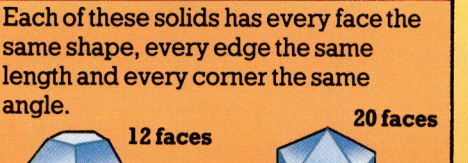

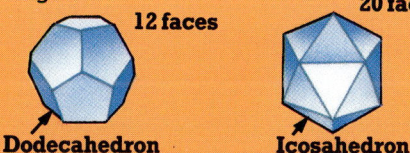

Each of these solids has every face the same shape, every edge the same length and every corner the same angle.

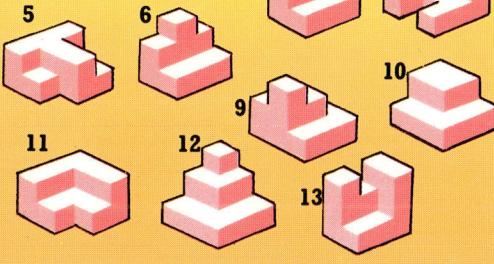

12 faces — Dodecahedron

20 faces — Icosahedron

*Hint – area of a circle = πr^2

Angles

Most people think of an angle only as a corner between two straight lines. In fact, angles often describe turns, like those of a tree being felled or a key in a lock. You can measure the size of an angle, that is, how many degrees (written °) the turn is, by using a protractor or angle indicator.*

Right angles

The corners of squares and rectangles are special angles called right angles. A full turn is 360° and four right angles make a complete turn. So a right angle is always 90°.

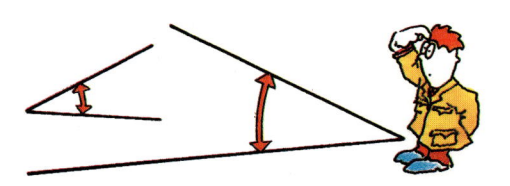

These angles are both 30°, so the length of the arms does not alter the amount of turn.

Angle names

A right angle is usually marked with a corner sign.

Angles which are less than 90° are acute.

Reflex angles are those which are more than 180°.

Obtuse angles are those between 90° and 180°.

Calculating angles without protractors

You can often work out the size of an angle by using logic rather than by measuring it. Logic enables you to arrive at theorems; a theorem is a mathematical rule which can be proved to be true. Below are some basic theorems involving angles.

1. Angles of a triangle

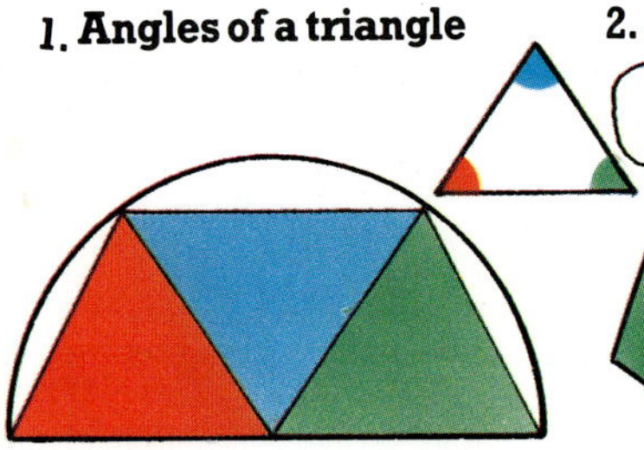

Tear off the corners of a triangle and place them together to make half a circle. This shows that the angles of a triangle together total 180°.

Because this is true for any triangle, many theorems of geometry are based upon this fact.

2. Angles inside polygons

180 × 4 = 720

Because any polygon can be divided into a number of triangles, you can always calculate the total value of its angles. For example, a hexagon will make four triangles, so the angles in a hexagon must total 720°. How many degrees is each angle in a regular hexagon? **

3. Angles on a straight line

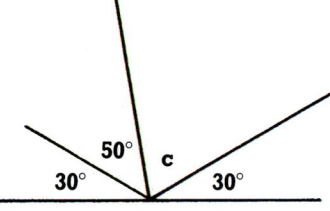

When angles fit together to make a straight line, you can imagine a half circle where they meet. So the sum of the angles on the line must be 180°.

This means that, if only one angle is unknown, it is possible to calculate it. Can you find the value of c?

*Go to page 37 to see how to use a protractor.
**Go to page 12 for more about regular shapes.

More angle theorems

Below, there are two more angle theorems. Because they refer to a number of angles, it is necessary to label the angles. An angle sign can be placed over the appropriate letter, for example, $\hat{B}$ or $\hat{E}$; or before the letter, e.g. $\lfloor B$ or $\lfloor E$.

In the picture on the right there are three angles meeting at D and you can define each one by showing the lines which form it.

A$\hat{D}$B, B$\hat{D}$A, A$\hat{D}$C and $\lfloor$CDA all describe the angle marked in red.

Alternate angles theorem

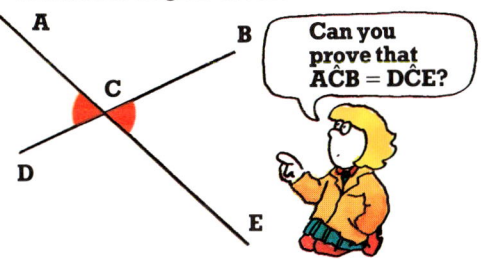

Can you prove that A$\hat{C}$B = D$\hat{C}$E?

Opposite angles are always equal.
Proof:
Where two lines cross, the four angles total 360°:
  A$\hat{C}$D + A$\hat{C}$B = 180° because DCB is a straight line.
  E$\hat{C}$B + A$\hat{C}$B = 180° because ACE is also a straight line.
  So A$\hat{C}$D = E$\hat{C}$B.

Exterior angles theorem

Use the exterior angle theorem to prove that D$\hat{E}$G = $\hat{A}$ + $\hat{D}$.

An exterior angle (such as F$\hat{D}$E) is equal to the two opposite angles of the triangle ($\hat{A}$ and $\hat{E}$).
Proof:
  $\hat{A}$ + $\hat{E}$ + A$\hat{D}$E = 180° because the three angles form a triangle.
  F$\hat{D}$E + A$\hat{D}$E = 180° because they form the straight line ADF.
  So $\hat{A}$ + $\hat{E}$ = F$\hat{D}$E.

Measuring in degrees

The first serious astronomers were the ancient Babylonians who believed that the year was made up of 12 moons each lasting 30 days and so there were 360 days in a year. It was perhaps for this reason that the circle was divided into 360 degrees. After 4000 years we still use these degrees to measure directions and angles.

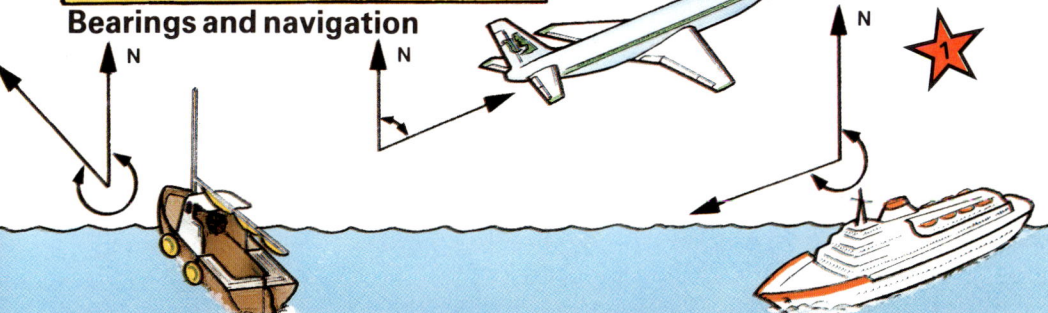

Bearings and navigation

Ships and planes record their direction of travel using bearings. A bearing is the angle measured clockwise from north. The aeroplane, barge and cruiser are travelling on bearings of 68°, 317° and 251°. Which bearing relates to which craft?

21

Topology

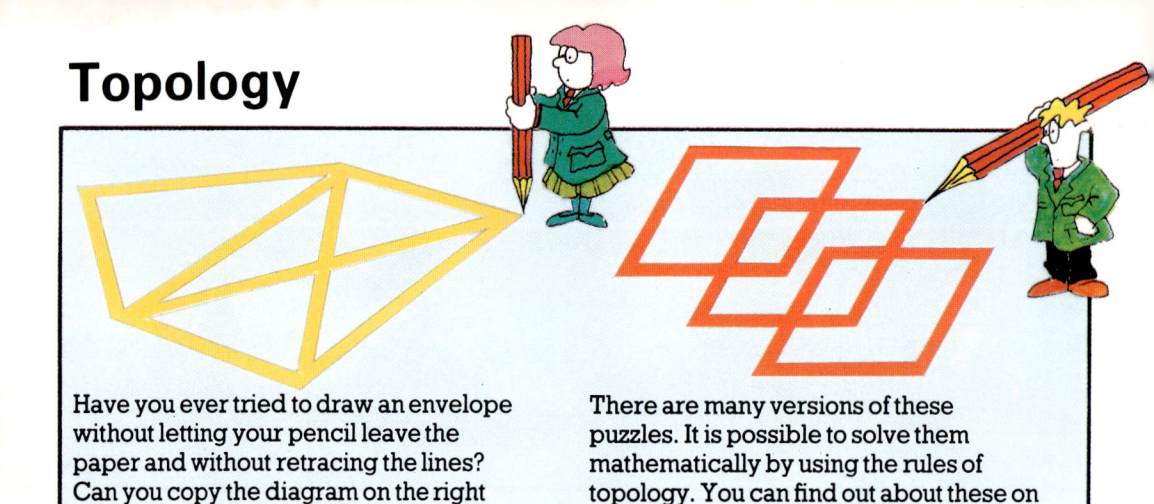

Have you ever tried to draw an envelope without letting your pencil leave the paper and without retracing the lines? Can you copy the diagram on the right without tracing the same line twice?

There are many versions of these puzzles. It is possible to solve them mathematically by using the rules of topology. You can find out about these on these two pages.

The seven bridges of Königsberg

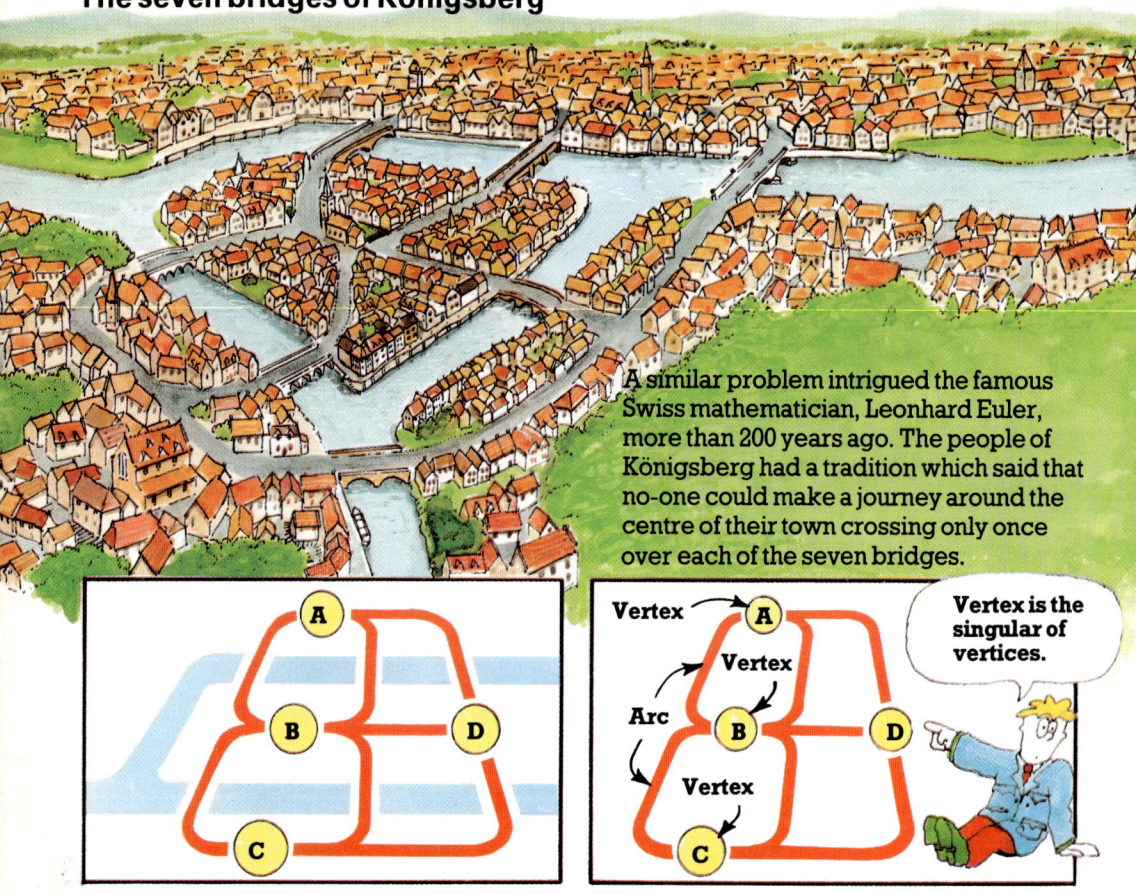

A similar problem intrigued the famous Swiss mathematician, Leonhard Euler, more than 200 years ago. The people of Königsberg had a tradition which said that no-one could make a journey around the centre of their town crossing only once over each of the seven bridges.

Euler drew a sketch map of the town, labelling the four separate parts A, B, C and D. He then marked each of the possible routes from one point to the next over the bridges, by tracing around the diagram with a pencil.

In doing this, he proved that it was not possible to cross every bridge just once. Euler called the diagram a network, the points vertices, and the lines arcs. If a network could be traced passing along all the arcs only once, he called it traversable.

Vertex is the singular of vertices.

Understanding networks

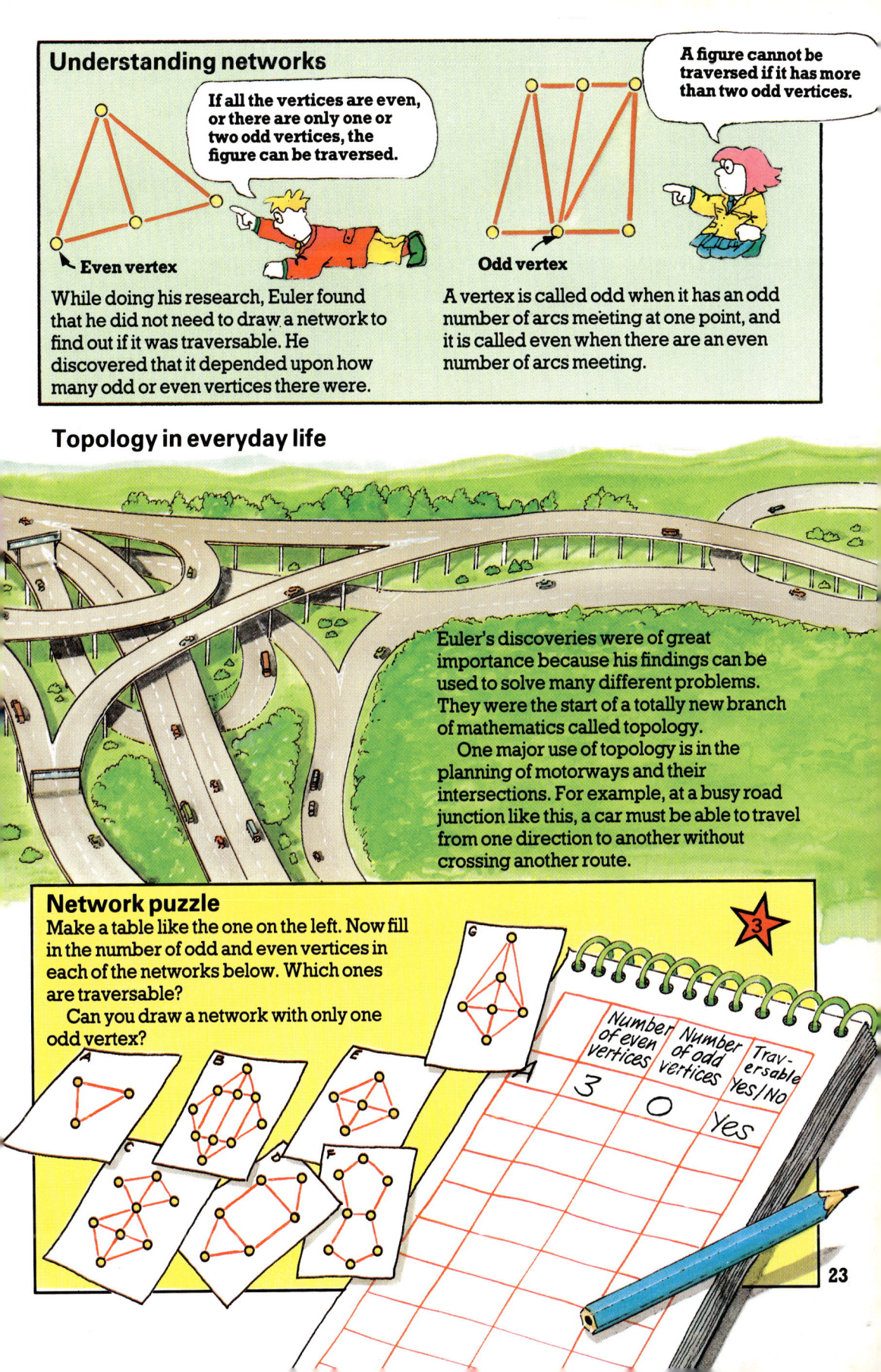

If all the vertices are even, or there are only one or two odd vertices, the figure can be traversed.

A figure cannot be traversed if it has more than two odd vertices.

Even vertex

Odd vertex

While doing his research, Euler found that he did not need to draw a network to find out if it was traversable. He discovered that it depended upon how many odd or even vertices there were.

A vertex is called odd when it has an odd number of arcs meeting at one point, and it is called even when there are an even number of arcs meeting.

Topology in everyday life

Euler's discoveries were of great importance because his findings can be used to solve many different problems. They were the start of a totally new branch of mathematics called topology.

One major use of topology is in the planning of motorways and their intersections. For example, at a busy road junction like this, a car must be able to travel from one direction to another without crossing another route.

Network puzzle

Make a table like the one on the left. Now fill in the number of odd and even vertices in each of the networks below. Which ones are traversable?

Can you draw a network with only one odd vertex?

Number of even vertices	Number of odd vertices	Traversable Yes/No
3	0	Yes

Fractions

You're ¾hr late!
My bike got a puncture when I was half-way here.

⅖ means 2 × ⅕.

We use fractions frequently in conversation without even realizing it. It is easy to picture the amount a fraction represents because it compares itself with 1.

This is ⅕ of the cake because 5 pieces like it make up the whole cake.

Sizing up fractions

There are two parts to a fraction, the numerator and denominator.

The larger the denominator, the smaller the pieces.

The numerator is the number of pieces.
2/5
The denominator shows the size of the pieces.
Which is bigger, 1⁄10 or ⅓?
1
2
3

Percentages

Percentages are fractions of 100. See if you can work out the percentages in this story.*

Astor has spent the week mending lunarbugs and has just been paid for his work. He goes into town to spend his money and his first stop is the bank.

Aries
20% off marked prices
240
BANK
Interest rate 10%

The bank will pay him an extra 10⁄100 or 1⁄10 of the money he deposits with them.

To calculate the interest on 120 solars at the rate of 10% per year, he multiplies 120 by 10⁄100.

1. How much interest will he have after one year?

2. How much money will he have in his account after 2 years?

Astor's next stop is Mr Aries' clothes shop. Mr Aries is offering a reduction of 20% – or ⅕ off (20⁄100 = ⅕).

To calculate the saving on this space suit, Astor multiplies 240 by ⅕.

3. How much will he save?

He could take the amount he saves away from 240 solars to obtain the price, but a quicker method would be to calculate 80% directly.

4. What is 240 × 80%?

*There are more puzzles involving percentages on pages 116-117.

Simplifying fractions

24/32 = 12/16 = 6/8 = 3/4

It is often convenient to change a fraction's denominator to make a calculation easier.

You can simplify a fraction by either multiplying or dividing both the numerator and denominator by the same number.

Adding fractions

To add fractions, first change each fraction to the same size sections, using the rules for simplifying fractions described above.

½ + ⅓ = ?
3/6 + 2/6 = 5/6

Both ½ and ⅓ can be changed to sixths – remember to change the numerators too.

Half of a half

½ × ½ = ¼

You might expect that multiplying would always make a number bigger – but not if you are multiplying by a fraction.

20 × ¼ = 20/4
20/4 = 5

3/8 × 2/3 = 6/24
6/24 = ¼

Subtracting fractions

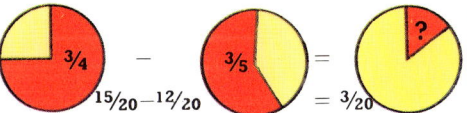

¾ − 3/5 = ?
15/20 − 12/20 = 3/20

These fractions can be changed to twentieths.

To multiply fractions, first multiply the numerators and denominators, then simplify.

Astor now decides to treat himself to dinner. When the time comes to settle the bill, he has to pay for the food, plus a service charge of 10% for the waiter, and a 15% cover charge to the proprietor.

The two bills drawn up by the proprietor and the waiter result in the same amount of money being owed by Astor.

The waiter would like to have the service charge added last, but the owner insists that the cover charge is the last to be added. You can see why!

BILL
26.00
3.90
+ cover 15% 29.90
2.99
+ service 10% 32.89
TOTAL

BILL
26.0
+ service 10% 2.6
28.6
+ cover 15% 4.29
TOTAL 32.89

Ratios

Ratios are to do with proportions. For example, if you make a jug of orange juice with two cups of concentrated orange and five cups of water, you are mixing the ingredients in the ratio of two to five (written 2:5). You can increase or decrease all the ingredients as long as the proportions stay the same.

You could make two jugs with four cups of orange and ten cups of water . . .

. . . or half a jug with one cup of orange and two and a half cups of water.

Working out ratios

1. Calculate the total number of shares.
2. Find the value of each share.
3. Find out how much each person gets.

There are 10 shares (5 + 3 + 2).

Each share is 180 ÷ 10 = 18 pieces.

Roy gets 5 × 18 = 90 pieces, Alison gets 3 × 18 = 54 pieces, I get 2 × 18 = 36 pieces.

Alison, Roy and Harry have just robbed the safe of a large bank and are trying to divide up the loot (180 pieces of gold) fairly. Since Roy is the leader, he wants a larger share than Alison or Harry.

Alison thinks that as Harry was only the look-out man, she should get more than him. They finally decide to share it out in the ratio of 5:3:2. To do this they have to follow the three steps displayed above.

Comparing ratios

Which material contains the most wool?

You may find it easier to compare these two ratios if you write the ratio 55:45 in the simpler form of 11:9.

VIYELLA
55% wool
45% cotton

CLYDELLA
2 parts wool
8 parts cotton

This tells you that the Viyella material contains 11 parts wool out of 20 parts in total. The Clydella material contains 4 out of 20 parts wool, so Viyella has more wool in it.

This map uses a scale of 1:25 000 which means that 1cm (or 1 inch) on the map represents 25 000cm (or 25 000 inches) on the ground. What would 1m on the same map represent?

Ratios are also used by bookmakers. You pay one token to bet on the horse Joyful, and if he wins you win five tokens. What would you win if you bet one token on Terror?

Horse Odds
Royal 9-1
Terror 100-8
Joyful 5-1

Ratios and volumes

A toy car is built in the ratio of 1:20 to a life-size car. But 20 toy cars would not even fill the boot of a real car, let alone equal it in size. Why?

20 toy cars

20 toy cars

20 toy cars

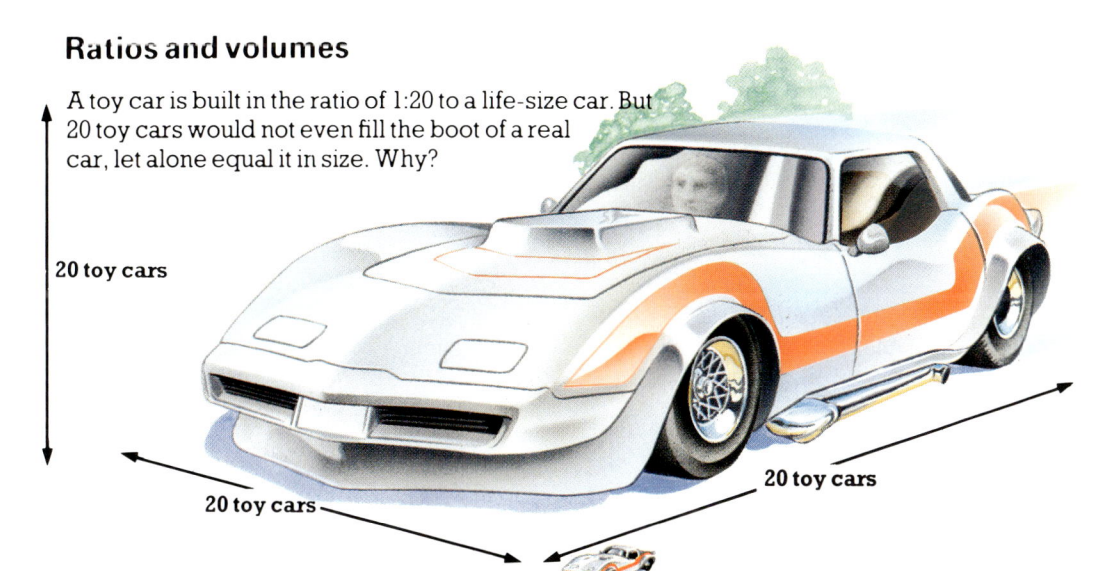

The answer is that the ratio of 1:20 gives a comparison of *each* length of the two cars. The toy car is 1/20th as long as the real one, 1/20th as wide and 1/20th as high.

You would need 20 × 20 × 20 toy cars to fill up the same space as a real car. So in fact, the ratio of the volumes of the two cars is 1:8000.

Puzzles

Bottle A holds 1 litre of wine and bottle B holds 2 litres. Why isn't bottle B twice the height of bottle A?

If it takes 50 biscuits to fill this jar, would a jar identical in shape but twice the size be filled by 100 biscuits?

Triangle ratios

Ancient Egyptians measured out the base of a pyramid with a rope which had 12 equal sections knotted on it. This was held taut in the shape of a triangle with sides in the ratio of 3:4:5. The angle opposite the longest side was always a right angle. Probably the most well-known theorem in mathematics, Pythagoras' theorem explains why.

Pythagoras' theorem

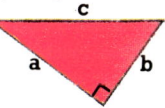

Pythagoras was a famous Greek mathematician. He discovered that when a triangle has a right angle and its three sides are, say, a, b and c, then $c^2 = a^2 + b^2$.

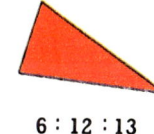

6 : 12 : 13   9 : 12 : 15   7 : 8 : 10

Above are the ratios of the sides of three different triangles. Only one of them has a right angle – which one? Use Pythagoras' theorem to find out.*

*See page 119 for more about Pythagoras' theorem.

Sets

Many different areas of mathematics involve organizing numbers and shapes into sets. Mathematicians have made it easier to study these sets by inventing the special symbols listed below.

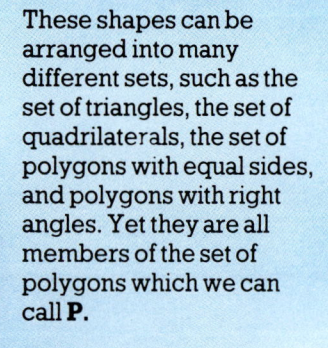

These shapes can be arranged into many different sets, such as the set of triangles, the set of quadrilaterals, the set of polygons with equal sides, and polygons with right angles. Yet they are all members of the set of polygons which we can call **P**.

If **A** is the set of those shapes with equal sides, then **A** = {b,d,j}.

If **B** is the set of those shapes with a right angle, then **B** = {a,d,f,h}.

The language of sets

∈	belongs to
⊂	subset of
A	the set **A**
A′	the set of members not in **A**
∅	empty set
∩	intersection (and)
∪	union (or)
n	number of members in

Can you identify the sets **C** and **D** when **C** = {a,b,c,h,i} and **D** = {d,e,f,g,j}?

C⊂**P** means that the set **C** is a sub-set of the larger set **P**.

b∈**C** means that **b** is a member of **C**. If **b**∈**C**, then **b**∈**P** because **C** is a subset of **P**.

The symbol n is used to show how many members a group has.

What is n(**A**∩**B**)?
What about n(**A**∪**B**)?

{circle, square, hexagon, cube, rhombus}

Odd one out
What could be the odd one out in this set?

Venn diagrams

In the diagram on the opposite page, some of the sets overlap because a shape such as **d** belongs in sets **A, B** and **D**. In 1880, John Venn developed a simple device to show relationships between various sets, called a Venn diagram.

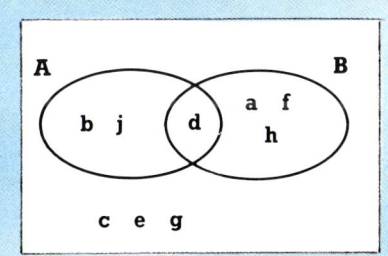

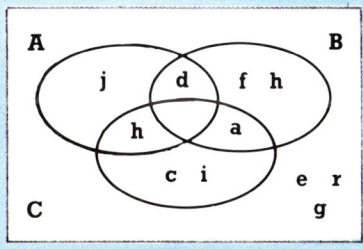

No member occurs twice.

This Venn diagram shows sets **A** and **B**. The overlap is **A ∩ B** and so **d ∈ A ∩ B**. Any element which is not a member of **A** or **B** is placed outside the loops.

This is a Venn diagram for sets **A, B** and **C**. The overlap of all three is **A ∩ B ∩ C**. The shape **a** has moved to be inside **B** and **C** but not in **A**, so **a ∈ A' ∩ B ∩ C**. What set is **A ∩ B ∩ C**?

Matrices

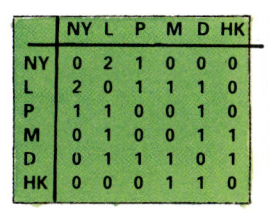

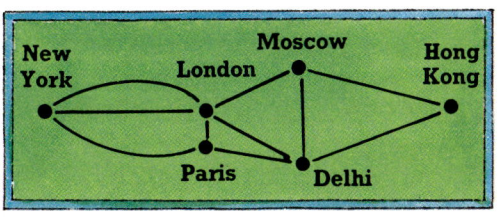

	NY	L	P	M	D	HK
NY	0	2	1	0	0	0
L	2	0	1	1	1	0
P	1	1	0	0	1	0
M	0	1	0	0	1	1
D	0	1	1	1	0	1
HK	0	0	0	1	1	0

Some information about sets of numbers is more easily stored in a matrix than in a list or Venn diagram.

The matrix above shows the number of direct flights between certain airports. While you can read the map just as easily for this small number of air routes, for a large amount of information a matrix is very useful.

Journey puzzle

Mrs Martin wants to travel from Hong Kong to New York but there are no direct flights. An airport official suggests a route she might take. Can you translate it?
HK ∩ D ∩ {L ∪ P} ∩ NY

Beyond numbers

$T = \{10, 11, 12, 13, 14 \ldots\}$

$Z = \{1, 2, 3, 4, 5, 6 \ldots\}$

Set Z is the set of counting numbers. It contains an infinite number of numbers which we can show by saying n(Z) = ∞.

Set T also contains an infinite number of numbers so n(T) = ∞. But T does not contain the first nine numbers so it looks as if ∞ − 9 = ∞.

This seems impossible to us unless we can accept the fact that we have argued correctly. The rules for numbers which describe the size of infinite sets seem to follow a different logic from ordinary numbers. It is an area that modern mathematics is still exploring . . .

Statistics

Statistics is a new branch of maths developed to record information and to help predict likely events. Statisticians obtain their information by collecting data, for example, by questioning a number of people who are representative of the population they want to test. This is called "sampling". Their findings can be presented in a number of ways, as shown below.

Researching a holiday resort

Lindsay works for a tour operator and is doing a survey on a holiday resort. She asks a sample of holiday makers various questions, such as "What nationality are you?", and "How did you travel to the resort?". She then displays her results in the five different forms shown below.

1. Pie chart

What is the most common nationality among the holidaymakers?

The circle below is divided up into sections which correspond to the percentage of each nationality.

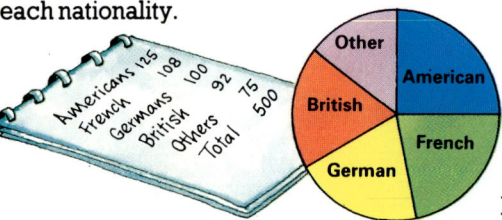

Americans 125
French 108
Germans 100
British 92
Others 75
Total 500

To calculate the angle for each section she finds out what fraction each nationality is of the total number of people. There are $^{125}/_{500}$ ($1/4$) Americans, so their angle is $1/4 \times 360° = 90°$. Can you calculate the other angles?

2. Block diagram

What method of transport did most people use to get to the resort?

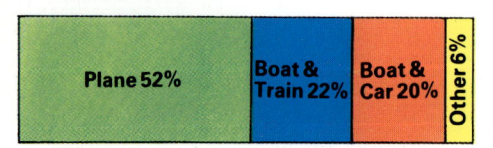

Plane 52% | Boat & Train 22% | Boat & Car 20% | Other 6%

This information is displayed in a block diagram. For a block diagram, a rectangle is divided up in the same way as the circle was divided up for the pie chart.

3. Histogram

How many people stayed at the Hotel Cana in the years 1978-84?

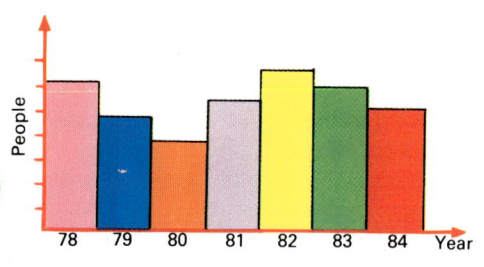

Each column of this histogram represents the number of people for that year; it is not related to any other year.

4. Frequency graph

How many people visited the beach in the first two weeks of June?

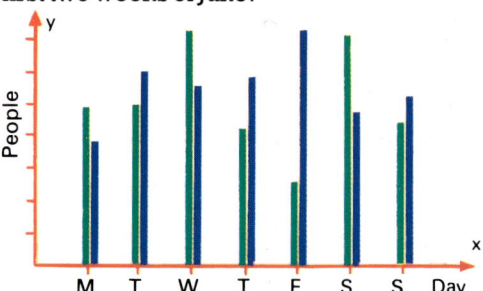

The advantage of this graph is that you can show a number of weeks on the same graph. Can you guess on which days it rained and on which days it was very hot?

Data compiled by statisticians is used by governments for long term planning, and by market researchers before launching a new product.

Three kinds of average

We often need to gain a quick understanding of the *general* size of a group of figures. Statisticians use one of three averages, the mode, median and mean, depending on the task.

Fred needs to know the average size of people's feet so that he can stock more sandals in that size. His kind of average is the mode: the most popular size.

The average size is 6.

5. Scatter graph

Is there a relationship between the amount of money spent and the age of the spenders? A scatter graph is used to plot information which seems unconnected, to see whether any trends appear.

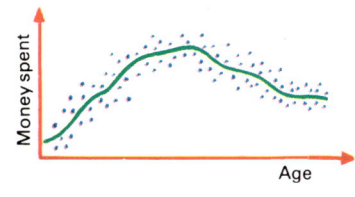

Here Lindsay plots the amount of money spent in a week against the age of the spender. What does the graph show?

In this game, the player must hit a weight with a hammer. If his hit is above average in strength, a bell rings. The maker of the game found the average strength by recording the scores of 100 people and finding the median. The median is the middle score.

The average most widely used is the mean. This is found by totalling all the items in the groups and dividing the total by the number of groups. Here the total number of crabs caught was 21. If they were shared out equally, each friend would get 7; so the mean caught was 7.

Statistics don't lie... but liars use statistics!

Look carefully at advertising or political statements which use statistics, as they can be misleading.

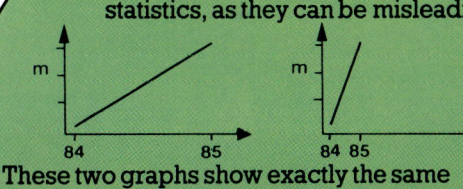

These two graphs show exactly the same information but the one on the right has been scaled to show a rise in the birth rate. Can you see how this has been achieved?

What are the mean, mode and median number of people who go swimming during this week?

Monday 380
Tuesday 415
Wednesday 380
Thursday 340
Friday 270
Saturday 300
Sunday 365

Graphs

Have you ever stopped to consider the relationship between the size of a bag of sugar and its weight? The larger the bag of sugar, the heavier it is. There are many different quantities which are related to one another in a similar way, such as the distance a car travels and its speed, or the age and height of a tree. The French mathematician, René Descartes (1596-1650) invented the co-ordinate system to illustrate such relationships on graphs.

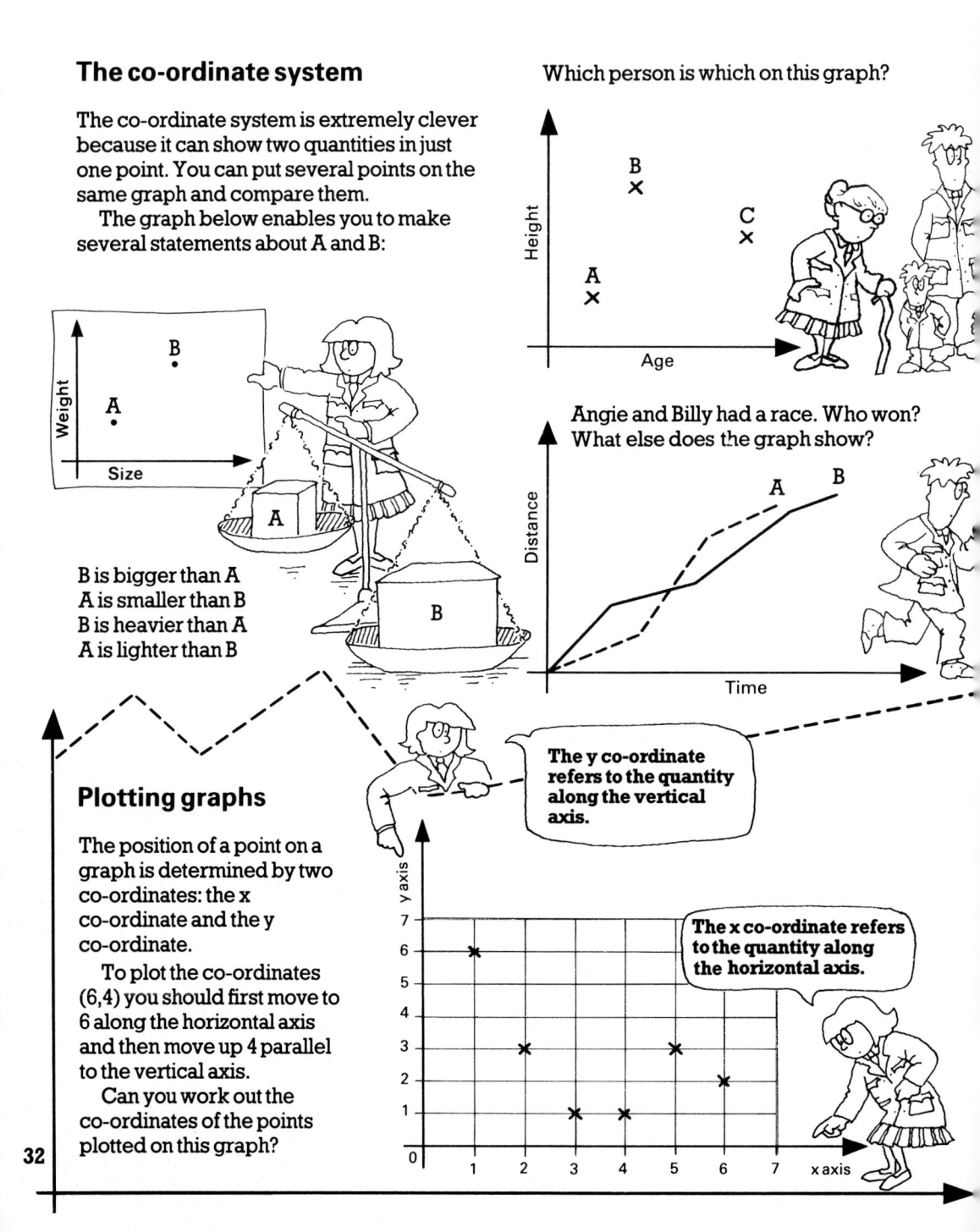

The co-ordinate system

The co-ordinate system is extremely clever because it can show two quantities in just one point. You can put several points on the same graph and compare them.

The graph below enables you to make several statements about A and B:

B is bigger than A
A is smaller than B
B is heavier than A
A is lighter than B

Which person is which on this graph?

Angie and Billy had a race. Who won? What else does the graph show?

Plotting graphs

The position of a point on a graph is determined by two co-ordinates: the x co-ordinate and the y co-ordinate.

To plot the co-ordinates (6,4) you should first move to 6 along the horizontal axis and then move up 4 parallel to the vertical axis.

Can you work out the co-ordinates of the points plotted on this graph?

The y co-ordinate refers to the quantity along the vertical axis.

The x co-ordinate refers to the quantity along the horizontal axis.

Picture puzzle

The pictures below tell the story of a hiker's journey. The graph on the right also shows her journey with each dot representing the distance travelled by the walker at a given time. Can you fill in the missing information in the pictures?

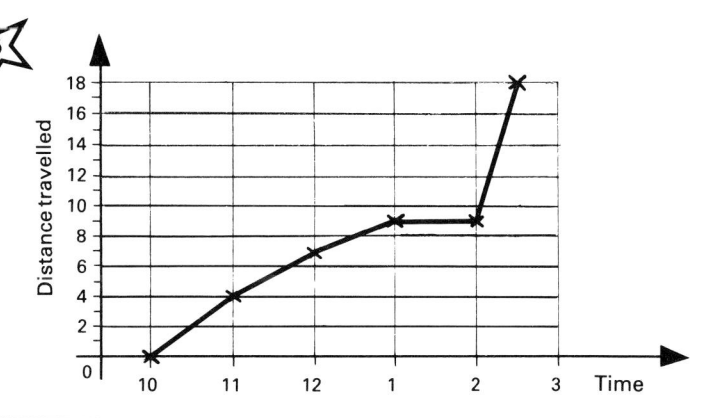

Reading graphs

The most important part of reading a graph is to notice what the axes are showing.

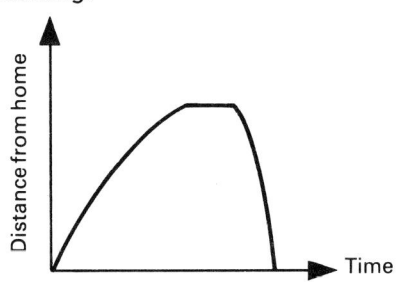

This graph shows the same journey as the graphs above, but the axes are labelled differently. Instead of showing the distance travelled in total, this graph indicates the distance the walker is from her starting point.

Warning

Don't jump to conclusions!

A graph may seem to indicate that, for example, people score more highly on a maths test as the shoe size gets bigger. Does that mean that the bigger your feet, the better you are at mathematics?

Always watch out for additional factors which may be causing the relationship.

More about graphs

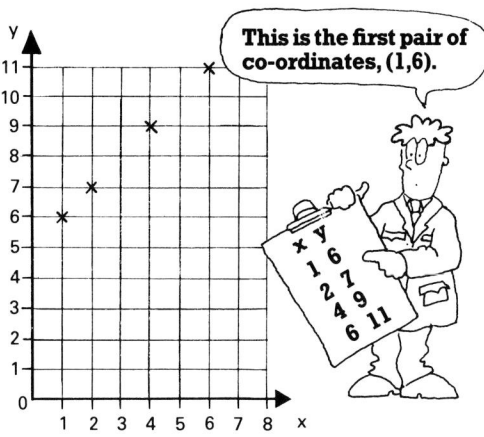

This is the first pair of co-ordinates, (1,6).

x	y
1	6
2	7
4	9
6	11

Look at the two sets of numbers on the pad above. Each number on the right is five more than the number on its left. The pairs of numbers are plotted on the graph above.

Because each pair of numbers has the same relationship (the y number is always five more than the x number), all the points lie on a straight line. Any other pair which has this relationship will also fit on the line, for example, ⁻2,3 and 2½,7½.

Describing relationships

You can write $y = x + 5$ to say that all the y numbers are five more than the x numbers. The line of dots on the graph is called the $y = x + 5$ line.

Another way of putting this is $x \rightarrow x + 5$. This is usually read as "x maps to x plus five". You can use whichever you find easier.

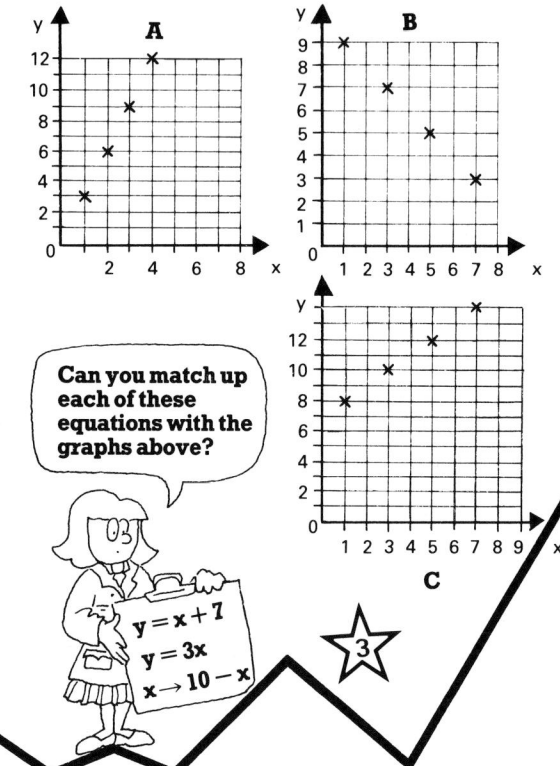

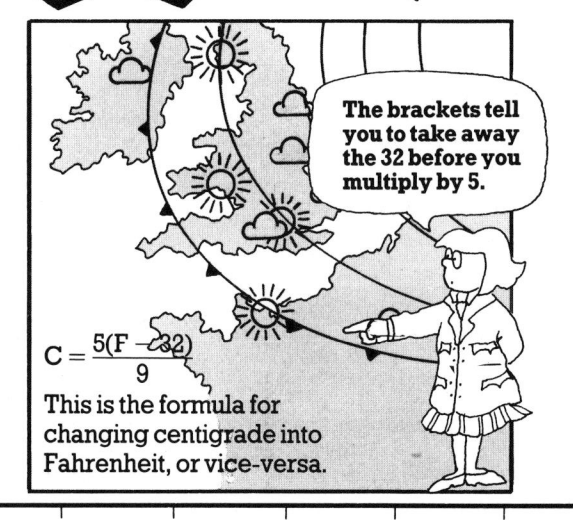

A

B

C

Can you match up each of these equations with the graphs above?

$y = x + 7$
$y = 3x$
$x \rightarrow 10 - x$

Solving equations

Each of the equations shown above describes a relationship which is true for many pairs of numbers. You could always find the value of y if you knew the value of x, and vice-versa.

On page 21 (Angles), you saw that $110 + c = 180$. Because the total value of the angles was 180°, you could find the value of c.

Another way to find c is to plot the graph for the equation $d = c + 110$.* You can then find the value of c when d is 180.

This method can be very time-consuming, and there are other ways of solving equations. On the opposite page are two different ways of solving these equations:

$2y - 7 = 11$ and $\dfrac{5(F - 32)}{9} = 20$.

The brackets tell you to take away the 32 before you multiply by 5.

$C = \dfrac{5(F - 32)}{9}$

This is the formula for changing centigrade into Fahrenheit, or vice-versa.

---

*You can use any letters you like to represent quantities in an equation.

Plotting equations

You can plot any relationship which is presented in either of the two forms described earlier. Follow the steps on the right to plot the graph $y = x + 3$.

1) Work out some number pairs which have this relationship:

x y
(1,4)
(3,6)
(5,8)

2) Plot the numbers on a graph.
3) Join up the points.

When working backwards, reverse the signs on the flags

Equation to be solved $2y - 7 = 11$

Function method

Explanation

$y \xrightarrow{\times 2} 2y \xrightarrow{-7} 2y - 7 = 11$

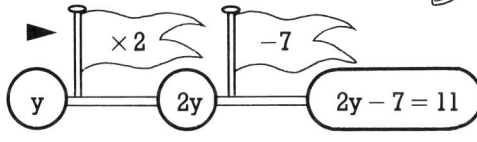

A value for y is doubled, then 7 is taken away from it. The answer is 11.

Solution

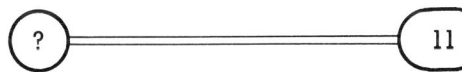

? ———— 11

To find out what value of y produces the answer 11, work backwards:

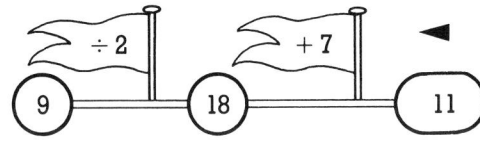

$9 \xleftarrow{\div 2} 18 \xleftarrow{+7} 11$

Answer: $y = 9$

Algebra method

Explanation

7 must be taken away from 2y to make the value 11.

Solution

If $2y - 7 = 11$
then $2y = 11 + 7$
so $2y = 18$
and $y = 9$.

Answer: $y = 9$

Equation to be solved $\dfrac{5(F - 32)}{9} = 20$

Function method

Explanation

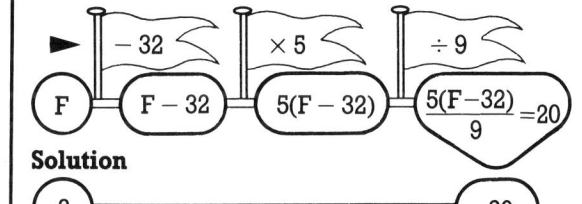

$F \xrightarrow{-32} F - 32 \xrightarrow{\times 5} 5(F-32) \xrightarrow{\div 9} \dfrac{5(F-32)}{9} = 20$

Solution

? ———— 20

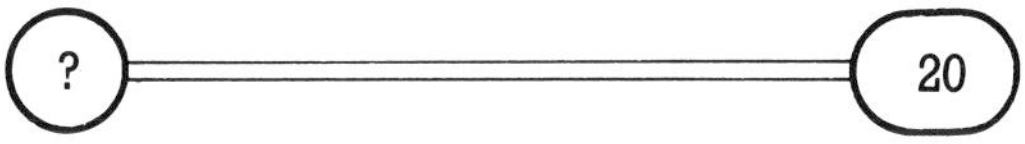

Again, you should work backwards:

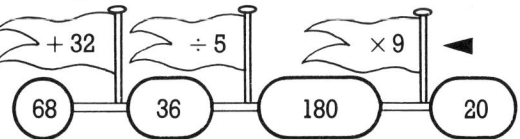

$68 \xleftarrow{+32} 36 \xleftarrow{\div 5} 180 \xleftarrow{\times 9} 20$

Answer: $F = 68$ so when the temperature is 20°C, it is 68°F.

Algebra method

Explanation

32 is taken away from F, and the number left is multiplied by 5, then divided by 9. The value after doing this is 20.

Solution

If $\dfrac{5(F - 32)}{9} = 20$

then $5(F - 32) = 20 \times 9$

so $5(F - 32) = 180$

and $F - 32 = \dfrac{180}{5} = 36$

If $F - 32 = 36$
then $F = 36 + 32 = 68$

Answer: $F = 68$

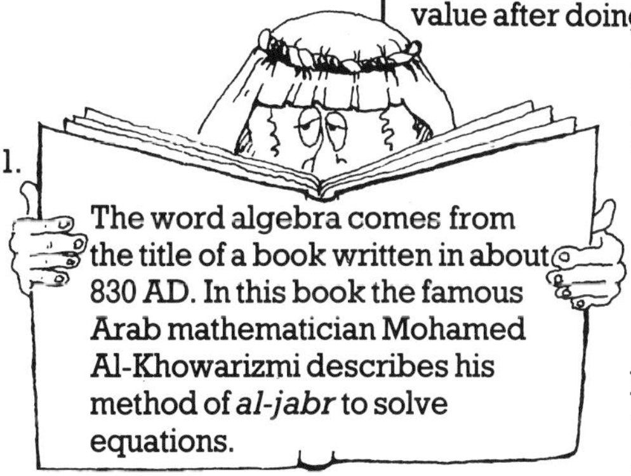

The word algebra comes from the title of a book written in about 830 AD. In this book the famous Arab mathematician Mohamed Al-Khowarizmi describes his method of al-jabr to solve equations.

35

Geometry

Geometry is to do with drawing accurate mathematical diagrams. Methods were being studied in Egypt as early as the 14th century BC when people paid their taxes according to the size of their land. Hence its name which comes from the Greek word *geo* (earth) and *metron* (measure). Today, map-making, surveying, aircraft design, architecture and computer circuitry all depend upon geometric precision. Below are some standard geometric constructions.

Make sure you have all the necessary equipment before you start.

Drawing a line of exact length

First draw a line longer than you want and make a mark near one end.
   Then open your compasses to the length you need.

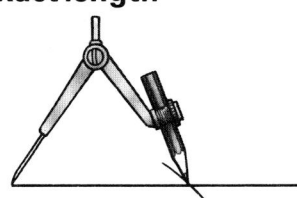

Place the compass point on the mark on the line and draw a light arc across the line with the pencil.

Drawing a triangle (sides 5cm, 3cm and 7cm)

1. Mark off a length of 7cm on a straight line.

2. Open the compasses to 3cm, place the point on one end of the line and draw an arc, as shown above.

3. Open the compasses to 5cm and draw a second arc from the other end of the line.

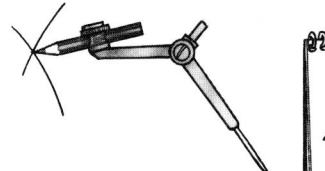

4. Join the three corners and check the measurements of the sides.

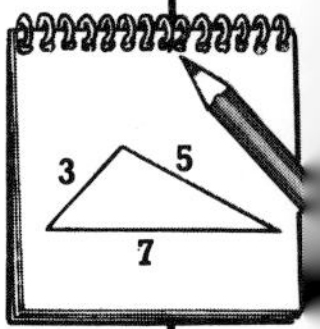

Greek mathematics today

Even today, the ancient Greeks are admired for the precision of their geometry. One of the most famous Greek mathematicians was Euclid. His painstakingly accurate diagrams enabled him to make some very surprising discoveries, two of which are described on the right.

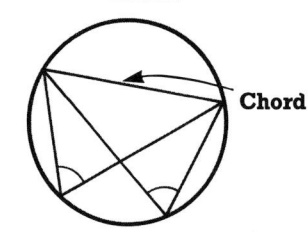

Chord

When a chord is drawn across a circle, any angle drawn from it to the edge of the circle will always be the same size, provided that the points are on the same side of the chord.

To bisect means to cut in half.

Bisecting a line

1. Open the compasses to more than half the length of the line.

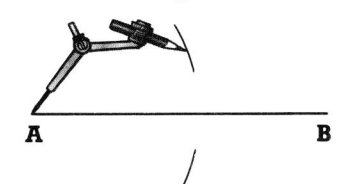

2. From one end of the line, draw an arc above and below the line.

3. Now do the same from the other end, without changing the compasses.

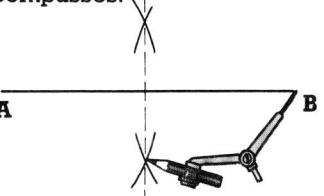

4. Join the two crosses and this line will cut AB exactly in half.

Bisecting an angle

1. Place the compass on each arc in turn, and draw two new arcs which cross each other, as shown below.

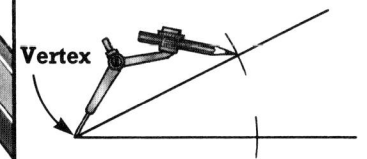

Vertex

2. Place the compass on each arc in turn, and draw two new arcs which cross each other, as shown below.

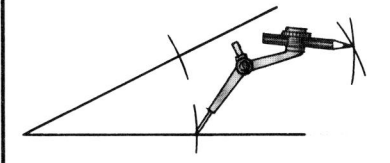

3. A line drawn from the angle to where the arcs cross will cut the angle in half.

Drawing parallel lines (2cm apart)

1. Draw a line about 4cm long.

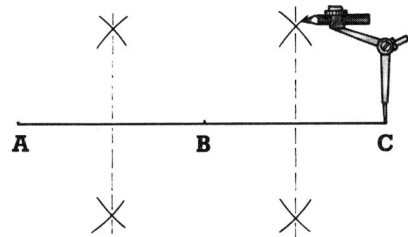

2. Mark three points, A, B and C. Bisect AB and bisect BC.

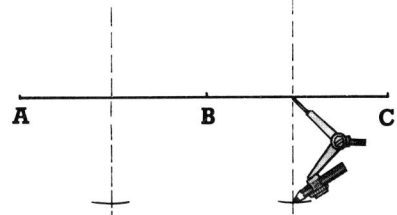

3. Mark off 2cm on each line. Join the two points where the arcs cross the lines.

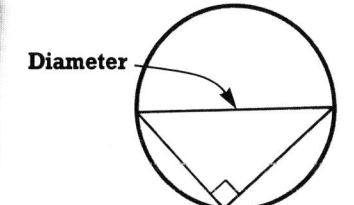

Diameter

When a chord goes through the centre of the circle (its diameter), any angle which you draw from its ends will be 90°, no matter where the point comes on the circle.

The easiest way to test these theorems is to draw some circles and chords, then measure the angles with a protractor.

1. Line up the zero line with one arm of the angle.

2. Place the centre of the protractor on the vertex of the angle.

3. Read off the size of the angle on the other arm. Is this 100° or 80°?

Number patterns

Since the time of Pythagoras (550 BC) mathematicians have made many of their discoveries by studying the patterns which occur in our number system. You can find out about a few of these patterns on these two pages.

Squares

The small figure ^{2} above any number tells you to multiply that number by itself.
$$2^2 = 4 \qquad 9^2 = 81 \qquad 12^2 = 144$$
The power 2 is usually called "square" because the multiplication can be illustrated by a square:

$3^2 = 9$

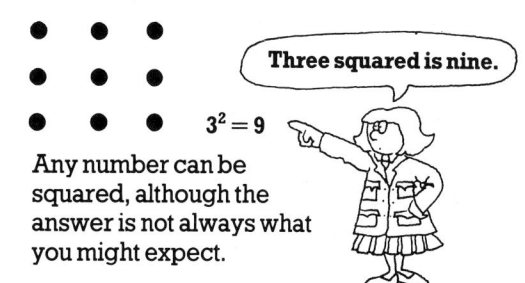

Three squared is nine.

Any number can be squared, although the answer is not always what you might expect.

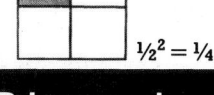

$½^2 = ¼$

Square roots

The square root of a number is the number which, when multiplied by itself, will give the original number.

The square root of 12.96 is 3.6 because 3.6×3.6 is 12.96.

Powers

Many calculations in mathematics involve multiplying numbers by themselves and it is quite common to meet powers greater than 2. 1 million is
$$10 \times 10 \times 10 \times 10 \times 10 \times 10 = 10^6$$ 
The power of 6 indicates that 6 tens are multiplied together.
$2^5 = 32$ because 2^5 means $2 \times 2 \times 2 \times 2 \times 2$.
The power of 3 is usually called cube.

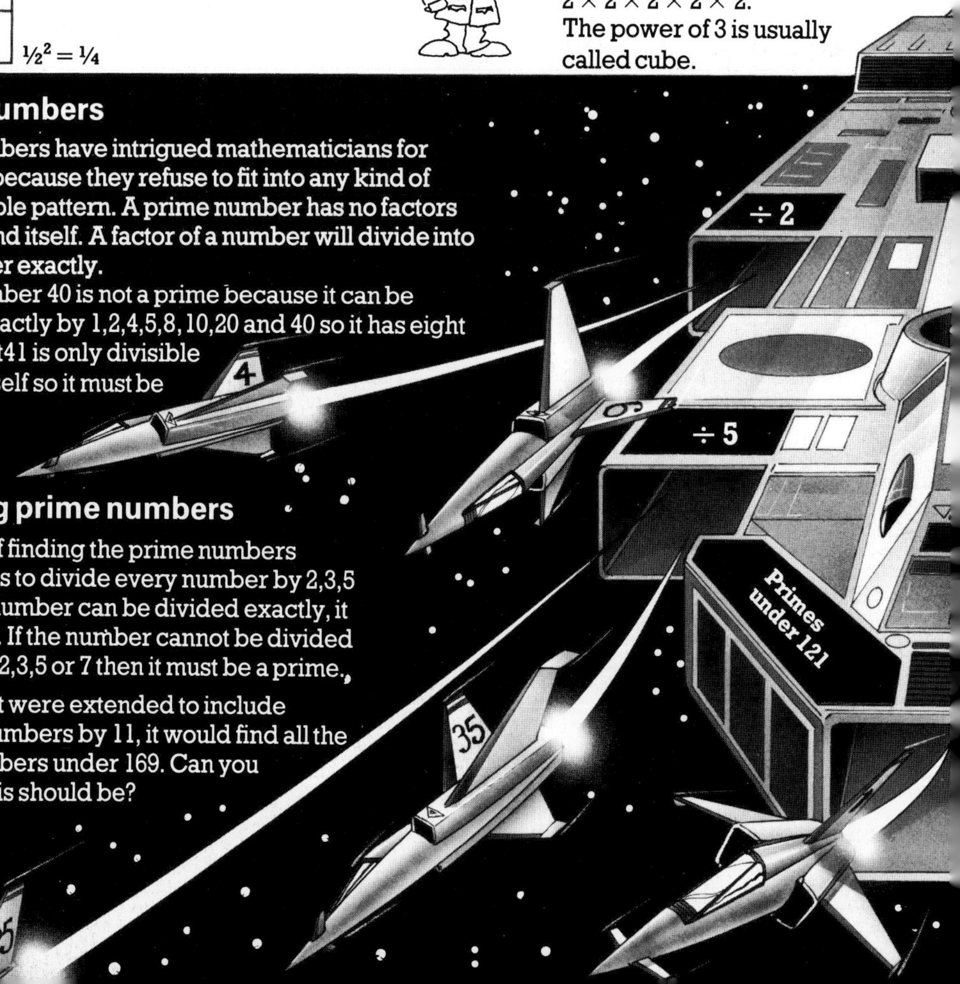

What is 2^3?

Another word for powers is indices.

Prime numbers

Prime numbers have intrigued mathematicians for centuries because they refuse to fit into any kind of recognizable pattern. A prime number has no factors except 1 and itself. A factor of a number will divide into that number exactly.

The number 40 is not a prime because it can be divided exactly by 1, 2, 4, 5, 8, 10, 20 and 40 so it has eight factors. But 41 is only divisible by 1 and itself so it must be a prime.

Spotting prime numbers

One way of finding the prime numbers under 121 is to divide every number by 2, 3, 5 and 7. If a number can be divided exactly, it is rejected. If the number cannot be divided exactly by 2, 3, 5 or 7 then it must be a prime.

If the test were extended to include dividing numbers by 11, it would find all the prime numbers under 169. Can you see why this should be?

Multiplying with powers

$2^5 \times 2^3 = (2 \times 2 \times 2 \times 2 \times 2) \times (2 \times 2 \times 2)$
$= 2^8$

When multiplying identical numbers with the same, or different powers, you need only add the powers.

What is $2^7 \times 2^2$?

Dividing with powers

$2^5 \div 2^3 = \dfrac{2 \times 2 \times 2 \times 2 \times 2}{2 \times 2 \times 2} = 2^2$

When dividing the same number with different powers, just subtract the indices.

$2^3 \div 2^5 = \dfrac{2 \times 2 \times 2}{2 \times 2 \times 2 \times 2 \times 2} = 2^{-2}$ or $\dfrac{1}{2^2}$

A negative power will always tell you that the number is a reciprocal. A reciprocal is a fraction where the top part is 1.

$10^{-3} = \dfrac{1}{10^3} = \dfrac{1}{1000}$

Can you write 10^{-5} in reciprocal form?

Standard form

Scientists frequently have to deal with very large numbers and very small numbers. For example, the distance from Earth to the nearest star, Alpha Centauri, is about 40 350 000 000 000 000 meters. The wavelength of sodium street lights is 0.000 000 589 meters. Because it is inconvenient to write numbers like this – they are hard to visualize and it is easy to make mistakes – they can be written in standard form.

Standard form is always a number between one and ten, multiplied by a power of ten. Because 40 350 000 000 000 000 is the same as $4.035 \times 10\,000\,000\,000\,000\,000$ it can be written as 4.035×10^{16}. Similarly, the wavelength of sodium light can be written as 5.89×10^{-7}.

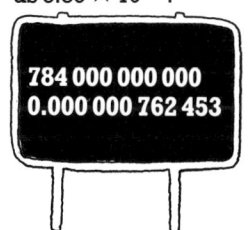

784 000 000 000
0.000 000 762 453

Can you write these numbers in standard form?

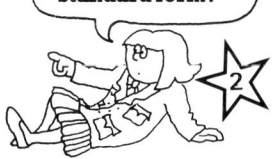

Formulae for primes

Many mathematicians have attempted to write formulae which would find prime numbers.

n^2 — n + 41 = p

Any number multiplied by itself. | Subtract the original number | Add 41 | The total is a prime number.

Although this formula works for some values of n, it fails when n is 41. Try the formula out and see for yourself.

Secret codes

No rules can be found to spot the large numbers which are primes; they can only be found by lengthy divisions. The most recently found prime number contains 25 692 digits and it took several weeks of computer time to produce it. Because prime numbers are so elusive, they are used in the most modern secret codes. It is the only use so far found for primes.

39

Binary numbers

Although computers and electronic calculators can cope with incredibly complex problems, they are unable to handle our decimal number system directly. They handle information in the form of patterns of electrical signals. Each signal can either be ON or OFF. Because of this, their number system is written using only two symbols: 1 (representing ON) and 0 (OFF). This is called the base two or binary system.

How binary works

The base two system works just like decimals (which is also called the base ten system). Each digit has a different value depending on its position in the total number. In the decimal system, the digits show the number of 1s, 10s, 100s, 1000s, etc. (see page 6). In the binary system, the digits show the number of 1s, 2s, 4s, 8s, etc.

So 23$_{base\ ten}$ is written 10111$_{base\ two}$.

Example: 23$_{base\ ten}$ in binary:

16s 8s 4s 2s 1s

1 0 1 1 1

$1 \times 16 \quad 0 \times 8 \quad 1 \times 4 \quad 1 \times 2 \quad 1 \times 1$

$16 + 4 + 2 + 1 = 23$

In binary you can write *any* number using just 1 and 0. To obtain the next column heading, just double the number which heads the column before it. Each column heading is, in fact, a power of two, and in decimals, each is a power of ten.*

To translate 37$_{base\ ten}$, for example, to binary, you need to work out how many 32s, 16s, 8s, 4s, 2s and 1s it contains. How would you write it in binary?

Translating binary

Example: 1011010$_{base\ two}$

64s 32s 16s 8 4s 2s 1s

1 0 1 1 0 1 0

$64 + 16 + 8 + 2 = 90$

So 1011010$_{base\ two}$ is 90$_{base\ ten}$. Can you translate 1100101$_{base\ two}$?

One less puzzle

Can you find the total of $1 + 2 + 3 + 8 + 16 + 32 + 64$ without doing any addition?

$1 + 2 = 3$
$1 + 2 + 4 = 7$
$1 + 2 + 4 + 8 = 15$
$1 + 2 + 4 + 8 + 16 = 31$
$1 + 2 + 4 + 8 + 16 + 32 = 63$

When you want to convert any number from binary to decimal, write the column headings above each digit starting on the right, and add the columns which have a figure 1.

*You can find out more about powers on pages 38-39.

Binary sums

It is possible to calculate in binary without translating the numbers into decimals.

The sum $1101_{base\ two}$ plus $101_{base\ two}$ is shown in separate stages below.

Two 1s make a 2 so carry it to the next column...

```
  8 4 2 1
  1 1 0 1
+   1 0 1
---------
        0
        1
```

```
  8 4 2 1
  1 1 0 1
+   1 0 1
---------
      1 0
```

Two 4s make an 8...

```
  8 4 2 1
  1 1 0 1
+   1 0 1
---------
    0 1 0
        1
```

and two 8s make a 16.

```
  8 4 2 1
  1 1 0 1
+   1 0 1
---------
  1 0 0 1 0
```

When subtracting one binary number from another you will often need to take 1 from 0.

...so take it from an 8 in the next column!

```
  8 4 2 1
  1 0 1 1
-   1 0 1
---------
        0
```

No problem so far!

```
  8 4 2 1
  1 0 1 1
-   1 0 1
---------
      1 0
```

You cannot take a 4 from no 4s...

```
  8 4 2 1
  1 0 1 1
-   1 0 1
---------
      1 0
```

```
  8 4 2 1
  1 0 1 1
-   1 0 1
---------
    1 1 0
```

Computer messages

When a computer sends a message to another computer using 1s and 0s we need to be sure that the receiving computer gets the right information. So a method is needed for detecting errors.

Why is this better than sending just 1 redundant figure for each digit?

One method is to send 2 redundant figures for each digit, so 1100111 would be 111 111 000 000 111 111 111. There are 2 transmission errors in the number 101000010111110000. Can you say what they are?

Other bases

You can in fact work in any base you wish, in base 4, for example, you could count in groups of 1s, 4s, 16s (4 × 4), 64s (4 × 4 × 4) and so on.

We still use base 60 for counting minutes and seconds. What is 1920 seconds in minutes?

Probability

The study of probability is a very exciting branch of maths because it allows us to gain some insight into the future. Although they cannot predict exactly what will happen, the laws of probability indicate what event is most likely to occur. The next three pages explain how to work out probabilities. On page 45 there are some programs to use on a home computer to find out how accurate the forecasts are.

Toss a coin in the air. It is equally likely that it will land Heads or Tails. So the chance of a coin coming up Heads is 50-50 or ½.

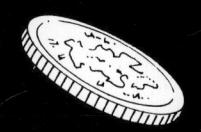

Dice Doubles

Many games involve throwing two dice. What is the probability of scoring a double?

To calculate the probability of something happening, first find out all the possible outcomes.

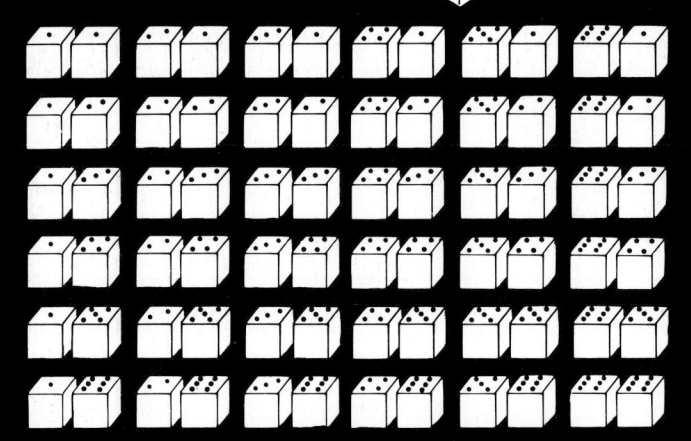

a) There are 36 possible ways in which the dice can land, all equally likely to happen.

b) Six of these possibilities are doubles so the probability of scoring a double is 6 out of 36 (or 1/6).

Go to page 45 to check the accuracy of these forecasts

Double doubles

What's the chance of throwing two doubles in a row?

Each time you throw a double with this first pair of dice, there is a one in six chance of throwing a double with the second pair. So there is a 1/6 × 1/6 (1/36) chance of throwing two doubles in a row.

Rule 1

When the outcome depends upon a series of events happening, multiply the separate probabilities.

Tree diagrams

For two children in a family to be boys, the probability is 1/4. This is because there are four possible arrangements for two children as this tree diagram shows, and a boy or girl are equally likely.

A tree diagram is a useful tool in probability because it enables all the possible arrangements to be found.

2 Boys   Boy and Girl   Girl and Boy   2 Girls

Winning chances

Use either of the methods below to work out your chances of winning at this game.

a) There are 52 playing cards which include 4 aces and 12 picture cards, i.e. 16 winning cards out of 52. The probability of winning is 16/52 = 4/13.

b) There are 4 aces, that's a probability of 4/52 (or 1/13) and 12 picture cards, another probability of 12/52 (3/13). That's a total probability of 1/13 + 3/13 = 4/13.

Pick an ace or a picture card to win.

Probability scale

Mathematicians use a scale of 0 to 1 to show the likelihood of an event. They often rely on statistics to judge whether an event is likely or not.

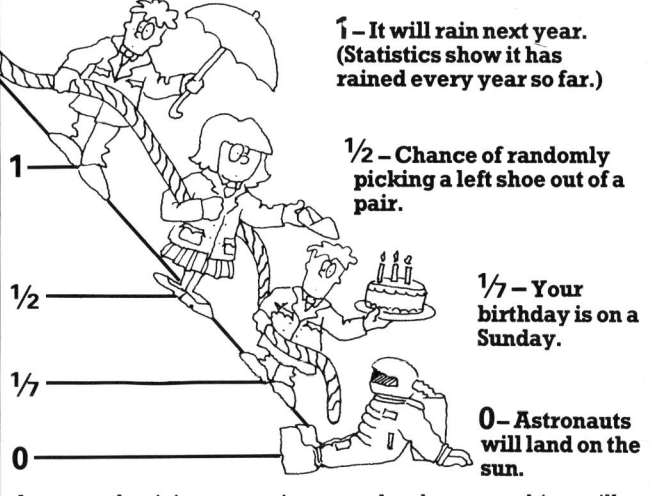

1 – It will rain next year. (Statistics show it has rained every year so far.)

1/2 – Chance of randomly picking a left shoe out of a pair.

1/7 – Your birthday is on a Sunday.

0 – Astronauts will land on the sun.

1 means that it is as certain as can be that something will happen, and 0 means that an event is so unlikely as to be thought impossible.

Rule 2

For mutually exclusive events, that is when an event is not dependant on any other, add the probabilities.

Bingo

In this Bingo game ten numbers have been called. What is the probability of making a complete line at the next call?

After ten numbers there are still another eighty-nine possible numbers for the next call. This player can complete a line if 25 or 33 are called. The probability of winning is therefore 2/89.

Matching birth signs

Did you know that in a group of three people there is a probability of 24/100 (24%) that two have been born under the same birth sign?

So the chance of two of us having the same birth sign is about 1/4 on the probability scale.

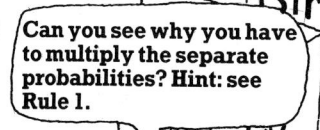

Can you see why you have to multiply the separate probabilities? Hint: see Rule 1.

You first need to work out the chances of them having different birth signs. Niki's sign is Pisces, and as there are 12 signs in all, there is a 11/12 chance that Fiona's sign is not Pisces. If they are different, there is a 10/12 chance that Alan will not match either of their signs.

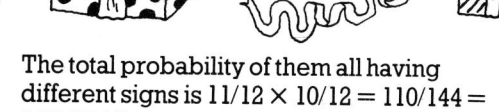

The total probability of them all having different signs is $11/12 \times 10/12 = 110/144 = 0.76$ (76%).

Quality Control

In a factory, thousands of lightbulbs are made each day. It is impossible to check each one, so a sample is taken.

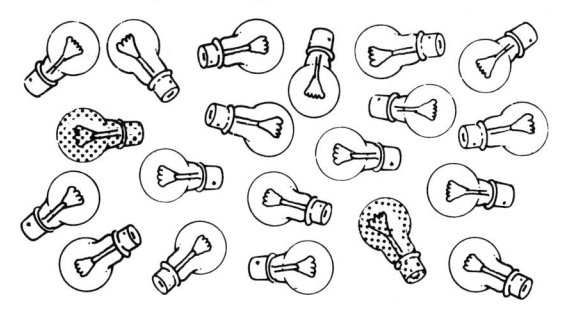

If one in ten of the sample lightbulbs are faulty, it is assumed that one in ten of all bulbs are likely to be faulty.

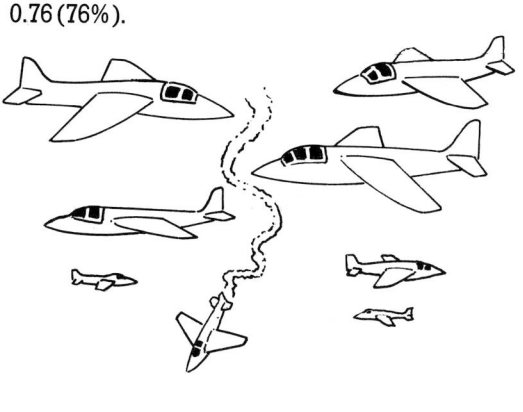

Although the probability of an error of 1 in 1000 may be acceptable in the case of lightbulbs, it would be disastrous as a gauge for aircraft as people's lives are at stake.

Computer programs

These programs will work on the Commodore 64, VIC 20, Apple, TRS-80 Colour Computer,* BBC, Electron and Spectrum. Lines marked ★ may need converting for the different computers. The conversions are listed at the bottom of the page. Run the programs and follow the instructions on the screen.

1. Prime number program

This program works out whether or not a number is a prime. You can use it to test out the systems on pages 38-39.

```
10 PRINT:PRINT "WHAT IS YOUR NUMBER"
20 LET F=0
30 INPUT N:IF N<3 THEN GOTO 70
40 FOR I=2 TO INT(N/2)
50 IF INT(N/I)=N/I THEN LET F=1:LET I=N-1
60 NEXT I
70 PRINT:PRINT N;" IS ";
80 IF F=1 THEN PRINT "NOT ";
90 PRINT "A PRIME NUMBER":GOTO 10
```

2. Dice program

This program makes the computer throw the equivalent of 36 pairs of dice and display the numbers on the screen. It provides a quick way to test the average number of doubles (see page 42).

```
★10 CLS
  20 PRINT:PRINT
  30 FOR I=1 TO 9
  40 FOR J=1 TO 4
★50 LET N1=INT(RND(1)*6+1)
★60 LET N2=INT(RND(1)*6+1)
★70 PRINT " ";N1;",";N2;
  80 IF N1=N2 THEN PRINT "*";
  90 IF N1<>N2 THEN PRINT " ";
 100 NEXT J
 110 PRINT:NEXT I:PRINT
 120 PRINT "PRESS RETURN TO CONTINUE"
 130 INPUT X$:GOTO 10
```

3. Fruit machine program

When you run this program, it asks you how many turns, or spins, you want. For each spin the names of three symbols come up in a row. Three identical symbols win a prize.

In each column of the fruit machine there are 10 cherries, 10 oranges, 6 lemons, 4 grapes, 2 melons and 1 bell. You can work out the probability of any combination being displayed using the methods on pages 42-44. The computer records how many times you win each prize. See if this relates to your probability forecasts.

```
  10 GOSUB 170:GOSUB 180
  20 PRINT "HOW MANY SPINS":INPUT NS
  30 FOR I=1 TO NS:FOR J=1 TO 3
★40 LET F=INT(RND(1)*33+1)
★50 LET X(J)=VAL(MID$(W$,F,1))
  60 NEXT J
  70 GOSUB 170:PRINT "SPINS : ";I
  80 PRINT:FOR J=1 TO 3
  90 PRINT F$(X(J));" ";:NEXT J:PRINT
 100 IF X(1)=X(2) AND X(2)=X(3) THEN LET
     N(X(1))=N(X(1))+1:PRINT "* PRIZE *"
 110 PRINT:PRINT "PRESS RETURN":INPUT X$
 120 NEXT I
 130 GOSUB 170:PRINT "TOTAL SPINS = ";NS
 140 PRINT:FOR I=1 TO 6
 150 PRINT "3 X ";F$(I);TAB(18);N(I)
 160 NEXT I:PRINT:STOP
★170 CLS:PRINT:PRINT:RETURN
★180 DIM F$(6):DIM N(6):DIM X(3)
 190 FOR I=1 TO 6:READ F$(I):NEXT I
 200 LET W$="111111111122222222223333334444556"
 210 RETURN
 220 DATA "CHERRY","ORANGE","LEMON"
 230 DATA "GRAPE","MELON","BELL"
```

Dice program conversions

Commodore 64:
```
10 PRINT CHR$(147)
```
VIC 20:
```
70 PRINT N1;CHR$(157);N2;CHR$(157);
```
Apple:
```
10 HOME
```
Spectrum:
```
50,60 Replace RND(1) with RND
```
TRS-80:
```
50,60 Replace RND(1) with RND(0)
70 PRINT N1;CHR$(8);N2;
```

Fruit machine conversions

Commodore 64 and VIC 20:
```
170 PRINT CHR$(147):PRINT:PRINT:RETURN
```
Apple:
```
170 HOME:PRINT:PRINT:RETURN
```
Spectrum:
```
40 LET F=INT(RND*33+1)
50 LET X(J)=VAL(W$(F))
180 DIM F$(6,8):DIM N(6):DIM X(3)
```
TRS-80:
```
40 LET F=INT(RND(0)*33+1)
```

*Extended BASIC version

Puzzle answers

Pages 4-5
How many?
The words which could accompany the numbers are as follows: Hours by aeroplane (1), hours by car (5), kilometers (200), miles (125).

Hundreds and thousands
1. 2233
2. 13
3. 900
4. The 2 on the left means 2000, and the 2 on the right means 2.

Pages 6-7
Brainteasers
0.8 is bigger than 0.396.
0.5, 0.50 and 0.500 all represent the same quantity. The extra noughts merely indicate that the number is exactly 0.5.

Rounding off
2400, 7.22

Dividing by decimals
Although it looks as though the decimal point moves, it is of course the numbers which change position.

Pages 8-9
Sums with negative numbers
$6 - 2 = 4$, $^-7 + 4 = ^-3$, $2 \times ^-3 = ^-6$, $^-4 \times ^-2 = 8$, and $1 \times 3 = 3$.

Vectors
Vector B is $\binom{-6}{4}$, Vector C is $\binom{8}{-3}$

Pages 10-11
2, 4 and 8
$128 \div 8 = 16$

Multiply by 9
Yes, 684 is divisible by 9
($6 + 8 + 4 = 18$ and $1 + 8 = 9$).

Multiply by 10
To multiply by 1000, add three noughts.

Multiply by 15
a) 120, b) 540, c) 1380

Beat the teacher
Write the sum down twice, forwards and backwards, and add downwards:

$\quad 1 + \quad 2 + \quad 3 + ... + \;998 + 999 + 1000$
$1000 + 999 + 998 + ... + \quad 3 + \quad 2 + \quad 1$

$1001 + 1001 + 1001 + ... + 1001 + 1001 + 1001$

Twice the sum of the number is $1000 \times 1001 = 1001000$ so the answer is 500500.

Pages 12-13
Classifying shapes
The building shown is called the Pentagon because it has five sides.

Regular polygons
A shape would need an infinite number of sides before it became a circle.

The family of triangles
It is not possible to draw a triangle whose sides are 5cm, 3cm and 9cm, because the length of the third side must be less than the sum of the other two sides.

Fitting shapes
The tessellation is made up of regular triangles, squares and hexagons.

Pages 14-15
Rectangular areas
$A_r = 1 \times w$ means that the area of a rectangle = length × width.

Sports centre puzzle
One way to find the area of this plan of a sports centre would be to divide it into two rectangles and a triangle. You could then use the relevant formulae to calculate the separate sections, and add them together.

Pages 17
If you sliced the cone straight down from its tip to the middle of its base, its cross-section would be a triangle (this is not, of course, a curve).

Pages 18-19
The two lines are the same length, but the arrow tips make the first line look longer.

Calculating volume
The shape on the left is $17cm^3$ and the shape on the right is $14cm^3$.
To find the cylinder's volume, multiply πr^2 by its height. So the formula is $\pi r^2 h$.

Building cubes
Shape 5 is left over.

Pages 20-21
Angles inside a polygon
A regular hexagon has six equal angles, so each angle would be $\frac{720}{6} = 120°$.

Angles on a straight line
$c = 180° - 110° = 70°$

Alternate angles theorem
$A\hat{C}B + A\hat{C}D = 180°$ because DCB is a straight line. $D\hat{C}E + A\hat{C}D = 180°$ because ACE is also a straight line. So $A\hat{C}B = D\hat{C}E$.

Exterior angle theorem
$\hat{E} + \hat{A} + A\hat{D}E = 180°$ because the three angles form a triangle. $\hat{E} + D\hat{E}G = 180°$ because they form the straight line AEG. So $D\hat{E}G = \hat{A} + A\hat{D}E$.

Bearings and navigation
The barge is travelling on a bearing of 317°, the cruiser on a bearing of 251°, and the plane on a bearing of 68°.

Page 23
Networks A, B, F and G are traversable.

Page 24
Sizing up fractions
$1/3$ is bigger than $1/10$th.

Percentages
1. Astor will have 12 solars interest after one year.
2. After one year, he has 132 solars, so after two years he has
$132 + \frac{132}{10} = 145.2$ solars
3. Astor will save 48 solars on the original price of the space suit.
4. $240 \times 80\% = 192$ solars.

Pages 26-27
Betting odds
If you put 8 tokens on Terror you would win 100 tokens if Terror came in first. So 1 token would win you $\frac{100}{8} = 12.5$ tokens.

Map scales
1m on the map would represent 2500m.

Puzzles
It is bottle B's volume which is twice the size of bottle A: its height is only one dimension.

No – you would get 8 ($2 \times 2 \times 2$) times the number of biscuits in the second jar: 400 biscuits.

Pythagoras' theorem
The triangle with sides 9, 12 and 15 has a right angle because $9^2 + 12^2 = 15^2$.

Pages 28-29
Set C = the set of those shapes with three sides (triangles).
Set D = the set of shapes which have four sides (quadrilaterals).
$A \cap B = \{d\}$; i.e. d is in both A and B so n ($A \cap B$) = 1.
$A \cup B = \{a,b,d,f,h,j\}$; i.e. the members which are in either A or B so n ($A \cup B$) = 6.

Odd one out
There are often several possibilities for an odd one out in a set. In this case, it could be the circle (the others have straight sides), or the cube (it is a three-dimensional shape).

Venn diagrams
$A \cap B \cap C = \emptyset$

Journey puzzle
Hong Kong and Delhi and either London or Paris, and New York.

Pages 30-31
Pie chart
The other angles, to the nearest whole number, are as follows:
French = $\frac{108}{500} \times 360° = 78°$
Germans = $\frac{100}{500} \times 360° = 72°$
British = $\frac{92}{500} \times 360° = 66°$
Other = $\frac{75}{500} \times 360° = 54°$

You can check that you have calculated correctly by adding the value of the sections up and checking that they total 360°.

Scatter graph
This graph shows that most money is spent by the people in the middle-age bracket.

Mean, mode and median
Mean = 350, Mode = 380, Median = 365.

Statistics don't lie . . .
The graph on the right has a smaller gap between 1984 and 1985 which makes it appear that the time lapse is less.

Pages 32-33
The co-ordinate system
The old lady is C, the child is A, and the man is B.
Angie won the race, but the graph also shows that Billy was leading early on in the race.

Plotting Graphs
The co-ordinates are (1,6), (2,3), (3,1), (4,1), (5,3) and (6,2).

Picture Puzzle
Picture 2: 11 o'clock, picture 3: 7, and picture 5: 9.

Pages 34-35
Describing relationships
A is y = 3x, B is 10 – x, C is x + 7

Pages 36-37
The angle shown is 100°.

Pages 38-39
Powers
$2^3 = 2 \times 2 \times 2 = 8$
$2^7 \times 2^2 = 2^9$
$10^{-5} = \frac{1}{10^5}$

Standard form
7.84×10^{11}
7.62453×10^{-7}

Spotting prime numbers
2, 3, 5, 7 and 11 divide into any number under 169 (or 13^2; the next prime number after 11), except a prime number. So you can use the first five prime numbers to find any prime number up to the square of the sixth prime number, and so on.

Pages 40-41
$37_{\text{base ten}}$ is $100101_{\text{base two}}$.
$1100101_{\text{base two}}$ is $101_{\text{base ten}}$.

One less puzzle
Each column heading in binary is one more than the sum of all the headings before it. So, in this case, double 64 and take 1 from it.

Computer messages
The number should read 111000111111000.
If you sent just one redundant figure, you would know when there was an error, but you would not know how to correct it.

Other bases
1920 seconds is 32 minutes.

Page 44
Matching birth signs
In this puzzle, the result depends upon a series of events occurring, so all the stages are multiplied together.

Score chart

50 or over	You're a genius!
35 or over	Well done – you're a good mathematician.
20 to 35	Average – if you think about things more carefully you can do even better.
Under 20	Never mind – take more care and your score should improve.

47

Maths words

Acute Angle smaller than 90°.
Algebra Rules used when working with equations.
Angle The space, or amount of turn between two straight lines.
Arc Part of the circumference of a circle.
Area The size of a shape's surface.
Axes Two perpendicular lines used to show scales when drawing graphs.
Bearing The angle of a line measured clockwise from North.
Binary A number system which uses only two digits, 1 and 0. Also called Base 2.
Bisect Cut in half.
Chord A line joining two points on the circumference of a circle.
Circumference The edge, or rim, of a circle.
Congruent Having identical sides and angles.
Co-ordinate The value of a point on a graph relating to the horizontal or vertical axis.
Degrees Units used to measure the size of an angle.
Denominator The bottom of a fraction.
Diameter A chord which goes through the centre of a circle.
Ellipse A regular oval, produced when a cone is sliced at an angle between a circle and a parabola.
Equilateral Having equal sides.
Factor A whole number which divides into another number exactly.
Geometry The drawing and study of accurate mathematical diagrams.
Graph A diagram showing the relationship between two different quantities.
Hyperbola The curve produced when a cone is sliced to make a larger angle with the base than that made by the sloping edge of the cone.
Indices Another word for powers (*sing.* index).
Infinite Never-ending.
Isosceles Triangle with two equal sides.
Mean Average produced by adding up a list of numbers and dividing by the number of numbers.
Median Middle number of a set of ordered numbers.
Mode Most frequent number in a set.

Network A mathematical diagram in which a number of points (vertices) are connected by lines (arcs).
Numerator Top part of a fraction.
Obtuse Angle between 90° and 180°.
Parabola Curve produced by cutting through a cone at an angle parallel to one of the sloping edges.
Parallel lines Straight lines which are always the same distance apart.
Percentage A fraction where the denominator is 100.
Polygon A closed, flat shape with straight sides.
Polyhedron A solid with flat surfaces.
Power A small number placed above and to the right of another number, telling you how many times to multiply the second number by itself. For example, $5^3 = 5 \times 5 \times 5$
Prime A number which has no factors except one and itself.
Pythagoras' theorem A theorem stating that the square of the longest side of a right-angled triangle is equal to the sum of the squares of its other two sides.
Radius Distance from a circle's centre to its circumference, i.e. half its diameter.
Ratio Proportion of one quantity to another.
Reciprocal A fraction where the numerator is one.
Reflex Angle between 180° and 360°.
Right angle 90°.
Scalene Triangle with each of its sides a different length.
Sector A slice of a circle.
Semi-circle Half a circle.
Similar Identical in shape.
Symbol A sign representing a quantity or idea.
Tangent Straight line touching the outside of a circle or other curve.
Tessellation Pattern made by placing several shapes side by side.
Traversable When a network can be drawn by passing over all the arcs only once, it is traversable.
Vector Quantities represented by arrows to show their direction and size.
Venn diagram A diagram which shows the relationship between various sets.
Volume The amount of space occupied by an object.

POCKET CALCULATORS

John Lewis
Edited by Helen Davies

Contents

50	Types of pocket calculators	74	Using the Pi key
52	A simple calculator	76	How to use a scientific calculator
54	Doing simple sums	78	Calculator logic
56	Car rally puzzle	80	Angles and triangles
58	Inside a calculator	82	Working out powers and roots
60	Calculator code	84	Very large and very small numbers
62	How a calculator works out sums	86	Simple statistics
63	Number tricks	88	Permutations
64	Using the memory	90	Statistics puzzle
66	Another kind of memory	92	Other keys
68	Working out percentages	93	Puzzle answers
70	Squares and square roots	96	Calculator power
72	Star warrior puzzle		

Illustrated by Graham Round, Graham Smith, Martin Newton, Gary Rees and Martin Salisbury

Designed by Graham Round and Kim Blundell

Mathematics consultant: Nigel Langdon

Types of pocket calculators

There are lots of different types of pocket calculators ranging from simple ones that just do basic sums to scientific calculators for doing complex mathematical calculations. There are also calculators which are specially designed to do certain jobs, for instance, those used by accountants and engineers, and those which airline pilots use for navigation.

On the next few pages you can find out how to do calculations on a simple calculator, then on pages 76-92 you can find out how to use a scientific calculator. There are lots of puzzles to solve and games to play to help improve your skill and accuracy.

Scientific calculator ▶

This kind of calculator has lots of special keys which work out complex calculations, such as sines, cosines and powers. Calculators with just a few scientific keys are sometimes called semi-scientific calculators.

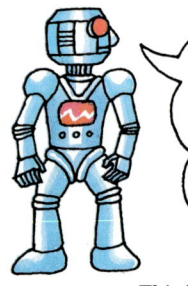

On/off switch

What is the largest number you can produce using the digits 1, 2, 3, 4, 5, and the x and = keys? You can only use each digit once. (Answer on page 93.)

This is a fortune-telling calculator. It has a calendar on it and to find out your fortune on any day you enter your date of birth, then press a special key marked "Birth".

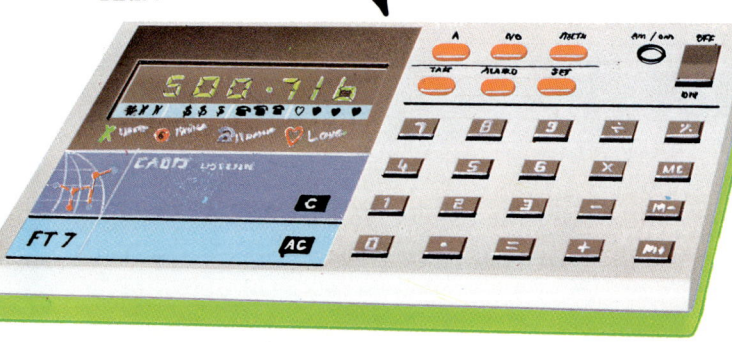

▲ Simple calculator

This is for doing everyday sums, such as adding or working out percentages. Simple calculators often have extra features, such as musical keys, or a game. Some have a built-in clock with an alarm.

Financial calculator ▲

Accountants and other business people often use a calculator like this. It has keys for working out interest on investments, depreciation, profits and losses.

▼ Programmable calculator

A programmable calculator can store the instructions for doing long, complex calculations and use them over and over again. A set of instructions is called a program. You can work out your own instructions and give them to the calculator by pressing the keys, or you can buy ready-made programs in small boxes called modules. These slot into the back of the calculator. Each module contains instructions for doing the calculations for a particular job, such as testing the strength of a bridge or navigating a boat.

These are the calculator keys.

This watch has a very small simple calculator built into it, and there is a game to play on the calculator too.

A module for a programmable calculator may contain 25 programs and up to 5 000 separate instructions.
You can buy modules with games programs.

Choosing a calculator

The first thing to decide is whether you want a simple or a scientific calculator. If you are buying a simple one think about which mathematical keys you may need. If you need to work out a lot of percentages, for example, make sure the calculator has a % key. Some calculators have a memory where you can store numbers while you are doing calculations. This is very useful. You may also want a clock or a game on your calculator. If you are a student, find out which keys you will need for your work and ask in the shops about calculators specially designed for students.

51

A simple calculator

This picture shows a simple calculator with all the keys marked to show the jobs they do. Each job is called a function. Adding, multiplying, squaring and finding percentages are all functions.

Your calculator may not look exactly like the one shown here. It may have fewer functions and some of the symbols on the keys may be different. If so, check the instruction booklet that came with it to find out what the symbols mean.

Display panel
Most simple calculators can show eight figures in the display panel, although they may use more in their calculations.

Can you work out how old you are in minutes?

Memory keys
These are for storing numbers in the calculator's memory. You can store the answer to one part of a sum while you work out the rest.

Percent key

Reciprocal key
You use this key to divide the number in the display into 1. This is called finding the reciprocal of the number.

Square key
Press this to square the number in the display, that is, to multiply it by itself.

Square root key
This key works out the square root of a number. The square root is the number which, when multiplied by itself would give the number you entered.

Pi key
This key puts the number 3.1415927 into the display. This number, called Pi after the Greek letter "π", is used for working out the circumference and area of circles.

Change sign key
This is for changing positive numbers into negative ones and negative numbers into positive ones. On some calculators it is labelled "CS".

Can you make the calculator display 50 by pressing only 7, 5, +, − and =? (Answer on page 93.)

Decimal point key

Equals key
Press this to finish a calculation. The answer then appears in the display.

On/off switch
Switching off the power clears all the numbers from inside the calculator.* To save batteries take care to switch off as soon as you have finished doing a sum. Some calculators switch off automatically after five minutes if no keys are pressed.

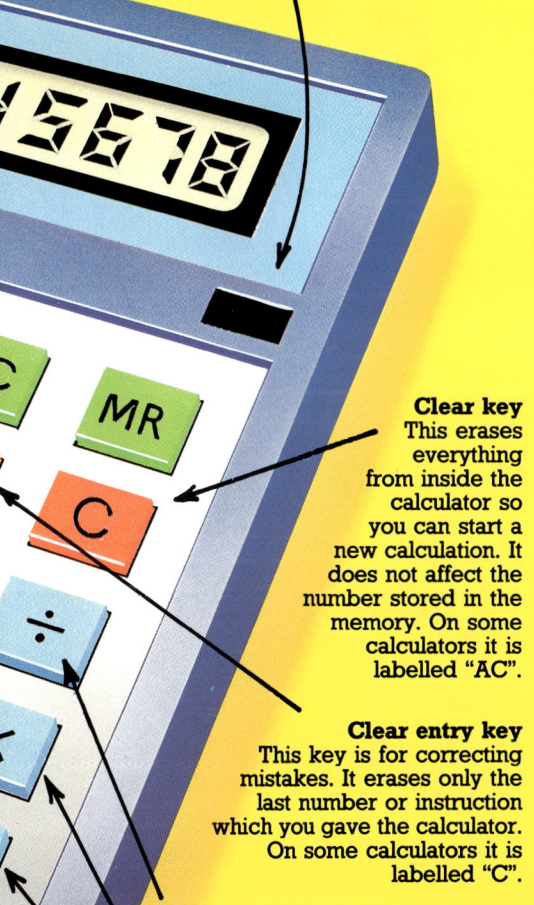

Clear key
This erases everything from inside the calculator so you can start a new calculation. It does not affect the number stored in the memory. On some calculators it is labelled "AC".

Clear entry key
This key is for correcting mistakes. It erases only the last number or instruction which you gave the calculator. On some calculators it is labelled "C".

Division key

Multiplication key

Subtraction key

Addition key

Pick several numbers between 1 and 9. Multiply each by 9, then by 12345679. What happens?

How the display works

Each digit in a calculator's display panel is made up using seven strips. The strips are arranged in a figure 8. Electrical signals from the calculator make the strips light up or change colour. By sending signals to various combinations of strips it is possible to make all the figures from 0 to 9.

In calculators with red displays the segments are made from tiny red lights called LEDs (light emitting diodes). In green displays they are made from a chemical which glows, like a fluorescent light. The black displays contain "liquid crystal" which turns black when it receives the electrical signal. Liquid crystal displays are the most common because they use least power.

Power

Most calculators use battery power. The batteries may be small torch batteries or special round ones which you can buy from calculator shops. You can also get rechargeable batteries and others which last for up to five years. Some calculators also have an adaptor and lead so they can be run from mains electricity.

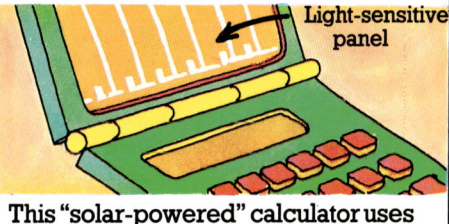

Light-sensitive panel

This "solar-powered" calculator uses light from the sun or an ordinary room light for power. When light shines on the panel it is converted into electrical power.

*On some memory calculators the memory is not cleared when you switch off.

Doing simple sums

Here are some sums to do using the functions +, −, × and ÷, to help you get the feel of a new calculator. When you put a number into a calculator it is called entering a number. For each sum you should enter the first number, press the key for the function, enter the second number, then press the = key. In divisions make sure you enter the number you are dividing into first. Before starting each sum press the clear key to clear the previous one from the calculator.

When you are doing sums watch the calculator's display. If you enter a wrong number you can correct it by pressing the clear entry key and then entering the correct number.

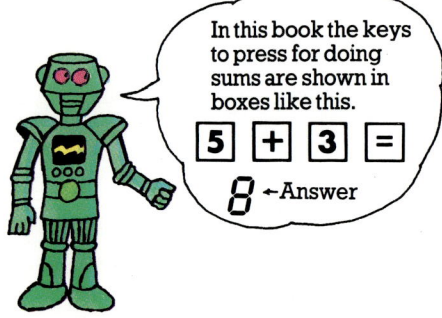

$1100 \times 9 + 9$

25×4

6059×423

The answers are on page 93.

$30 \div 2.4 - 3$

$9521 - 4193$

$10.4 \div 2$

$6400 \div 8$

$6 \times 10 \times 100$

$4098 - 3097$

$805.2 - 605.2$

$700 \times 4.3 \div 301$

$54 \div 4.5 - 6$

$9 + 17$

In this book the keys to press for doing sums are shown in boxes like this.

| 5 | + | 3 | = |

8 ← Answer

Negative numbers

| 6 | +/− | + | 4 | = | −2 |

Numbers such as −6 are called negative numbers because they are below zero. You can enter a negative number using the "−" key, or by pressing the change sign key (+/− or CS) after you have entered the number, as shown above.

Spotting mistakes

It is very easy to get the wrong answer on a calculator without realising it. You may enter a wrong number, or press the wrong function key. So it is a good idea to make a rough guess at the answer in your head to check the figure that the calculator gives you.

How to make a rough guess

1000 × 90 = 90 000

Suppose you do the sum 1 023 × 91 and the calculator shows the answer 930 930. To make a quick check, round off the numbers to the nearest ten, hundred or thousand. In this case change 1 023 to 1 000 and 91 to 90. Then, in your head, work out 1 000 × 90. The answer is 90 000, so the calculator's figure seems too big and you should enter the sum again. In fact, the correct answer is 93 093.

More sums

Try these examples on a calculator and do rough guesses to check your answers. The answers are given on page 93.

Figures for rough guesses

512 × 359	500 × 400
971 × 28	1000 × 30
1594 + 273	1600 + 300
6123 ÷ 57	6000 ÷ 60

Date of birth detector

Here is a trick for working out when someone was born. Ask a friend to enter into a calculator the day of the month on which they were born. Say their date of birth is 7 September 1970. They enter 7. Then you tell them to do the following:*

Multiply by 20, add 3 and multiply by 5.

7 × 2 0
+ 3 × 5

Add on the number of the month they were born, and again multiply by 20, add 3 and multiply by 5.

+ 9 × 2 0
+ 3 × 5

Then add on the last two figures of the year in which they were born.

+ 7 0 =

72485

To work out their date of birth, take the calculator and subtract 1 515 from the number in the display. Then if you read the figures in the display from left to right, you have the day, month and year in which they were born.

− 1 5 1 5
= 70970

*If you have a scientific calculator you may need to press = between each part of this calculation.

Car rally puzzle

Rally drivers often carry a calculator to work out how fast they should go over different sections of a race. In this puzzle you are the navigator in a rally car. The road conditions and your speed vary over the different sections of the race. Your job is to work out whether you can complete the race in the time allowed.

A

The total distance from A to the next check-point at D is 19.5km and you must be there in half an hour.

B

The first 9km from A to B are along a winding, hilly track and take 11.25 minutes.

You get stuck crossing a boggy field so your time over the next 7km from B to C is 16 minutes.

Rounding off answers*

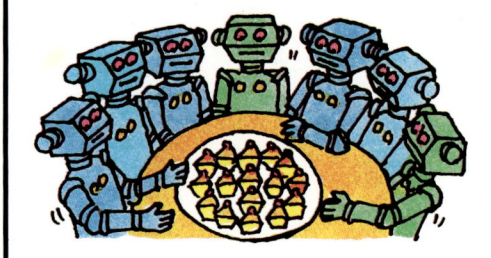

A calculator often gives a far more accurate answer than you need. For instance, if you had to divide 16 cakes amongst these 7 robots, a calculator would give the answer as 2.2857142 cakes for each robot.

It is impossible to cut cakes so accurately, so you do not need all the figures after the decimal point. Rather than just ignore them, though, you should "round off" the answer.

In most cases it is accurate enough to leave just one figure after the decimal point. To do this look at the second figure to the right of the point. If it is 5 or above, add 1 to the figure on its left. If it is 4 or below leave the figure on the left as it is.

In this example you can round the answer off to 2.3, so each robot would get about two and a third cakes.

*For more about rounding off, see page 7 and pages 110-111.

Then at C you take a wrong turning which adds an extra 1km to your journey. In addition, the road you are on has a speed limit of 100km/h.

Can you reach the check-point on time without breaking the speed limit?

How to work it out

First you need to work out how far you still have to travel from C to D. To do this subtract the distances from A to C from the total distance; then add on the extra 1km which the wrong turning caused.

Then you need to find out how much time you have left. You know how long it took to cover the first two sections. So how many minutes do you have left? You will need to change this to hours ($\div$ 60) to calculate the speed in km/h.

Now, by dividing the distance left (in km) by the time left (in hours) you can work out what speed you must do to reach the check-point in time. Will you break the speed limit? The answer is on page 93.

Remember the speed you work out is your average speed. You are unlikely to be able to travel at exactly this all the way. You will go below and above it.

Another car which set out from A at the same time as you reached the check-point 5 minutes before you. What was its average speed? (Answer on page 93).

Doing sums with fractions

Keys to press to convert ½: $1 \div 2 =$

$1/10$ $21^2/_3$ $9/8$
$15/16$
$5/17$
$3^5/_8$ $7^3/_4$

Most calculators cannot work in fractions so you have to convert them to decimals. To do this you divide the top part of the fraction by the bottom part.

To convert mixed numbers which have fractions and whole numbers (for example, $3^{11}/_{16}$), you only need to work out the fraction part because the whole number does not change. (11 $\div$ 16 = 0.6875 so $3^{11}/_{16}$ = 3.6875).

Can you convert these fractions to decimals? The answers are on page 93.

When you are doing sums you should not round off the numbers until the very end. You should always use all the figures after the decimal point for the calculations.

Inside a calculator

This picture shows the parts inside a calculator where the calculations are done. When you press the keys to enter the numbers and functions for a sum, electrical signals are sent inside the calculator and the numbers and functions are converted into a special code which the calculator uses for doing calculations. Below you can find out how the calculator does a simple sum.

① Scanners
When the calculator is on, these constantly search the keyboard, waiting to pick up the electrical signal made when a key is pressed.

② Encoder unit
This is where the numbers and functions are converted into the code which the calculator uses for doing sums.

④ Flag register
The function for the calculation is stored here until the calculator needs it.

⑥ Permanent memory
The instructions telling the calculator how to add, subtract, do percentages, square roots and all the other functions it can carry out are stored here. The instructions are called programs and they are put in the permanent memory when the calculator is made.

A simple sum

Imagine you are doing the sum 23 + 7. When you enter 23 it is picked up by scanners ①, coded ② and sent to the X register ③.

Next you press the + and this is also picked up by the scanners and coded. Then it is sent to the flag register ④.

The second number, 7, is coded and sent to the X register. This pushes the first number out into the Y register ⑤.

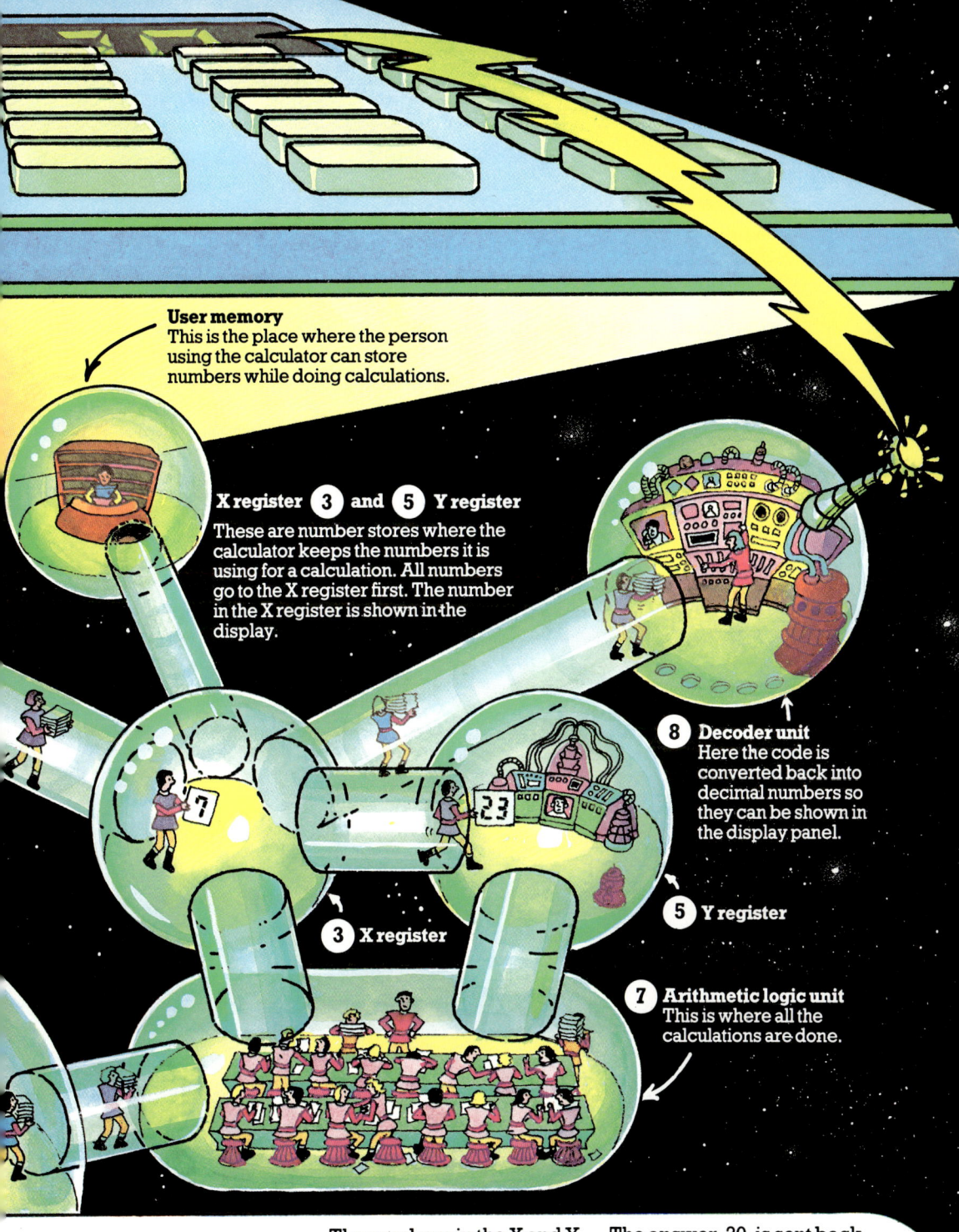

User memory
This is the place where the person using the calculator can store numbers while doing calculations.

X register ③ and ⑤ Y register
These are number stores where the calculator keeps the numbers it is using for a calculation. All numbers go to the X register first. The number in the X register is shown in the display.

⑧ Decoder unit
Here the code is converted back into decimal numbers so they can be shown in the display panel.

⑤ Y register

③ X register

⑦ Arithmetic logic unit
This is where all the calculations are done.

When you press = a message from the flag register tells the permanent memory ⑥ that the sum is an addition.

The numbers in the X and Y registers are then loaded into the arithmetic logic unit ⑦ and the calculation is carried out following instructions from the permanent memory.

The answer, 30, is sent back to the X register. From there it goes to the decoder unit ⑧. It is converted into a decimal number and then shown in the display.

59

Calculator code

The code which the calculator uses for doing sums is made up of pulses of electricity. There are two signals in the code: pulse and no-pulse. It is called a binary code* and can be expressed in numbers with a "0" for no pulse and a "1" for pulse. Different arrangements of the signals are used to represent the numbers, the functions and all the other information in a calculator.

No-pulse
Pulse

The patterns of signals are controlled by tiny electronic components called transistors. These act like gates, opening to let electric pulses through and shutting to stop them.

Number codes
The numbers we use (0-9) are in decimal code. Below you can see the difference between decimal and binary number codes.

Decimal code
In decimal code there are ten digits: 0123456789. When you write a number, each digit represents a set of units. The right-hand digit represents 1s. Each time you move to the left the size of the units increases ten times. So, in the number 3 507 there are seven 1s, no 10s, five 100s and three 1 000s.

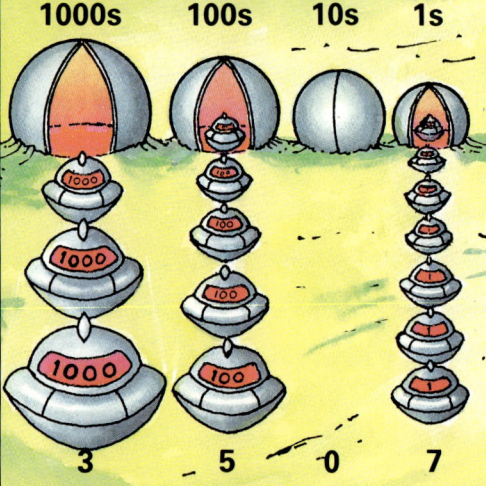

1000s   **100s**   **10s**   **1s**

3   5   0   7

3000 + 500 + 0 + 7
= 3507

Added together this makes three thousand, five hundred and seven.

More about transistors
Inside a calculator there are hundreds of thousands of transistors. The transistors are linked together in pathways, called circuits, along which the electric pulses travel. The circuits containing transistors are engraved by a special chemical process onto a tiny chip of a substance called silicon. The chip of silicon is called an integrated circuit, or just a "chip".

The chips of silicon are enclosed in plastic cases like the ones shown here. The metal legs carry the electric current from the battery into and out of the chip.

*See pages 40-41 for more about the binary code.

Binary code

In binary code there are only two digits. When you write a number the right-hand digit represents 1s and each time you move to the left the size of the units doubles. So in the binary number 1011 there is one 1, one 2, no 4s and one 8.

8s   4s   2s   1s

1   0   1   1

8 +   0 +   2 +   1
       = 11

Added together this gives the decimal number 11. So 1011 is 11 written in binary code.

Binary decoder

Here is an easy way to decode binary numbers into decimal ones.

Imagine you have a row of lightbulbs numbered 8, 4, 2, 1. If the switch below a bulb is down, the light is on and it represents a binary 1.

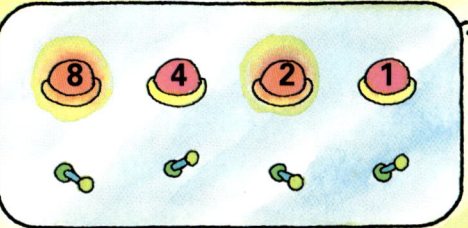

To decode a binary number, say 1010, start from the right of the binary number and for each binary 1 switch on the corresponding light on the decoder. Then add up the numbers on the bulbs that are lit to get the answer in decimal numbers. Can you decode the binary numbers shown below?

0011   1111   0101

0111   1100   1001

The answers are on page 93.

Actual size of chip

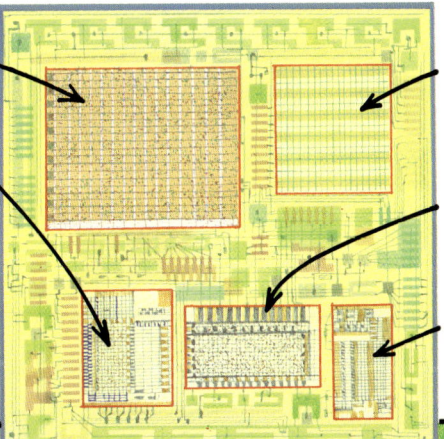

Permanent memory

Binary encoder unit

X, Y and flag registers

Arithmetic logic unit

Binary decoder unit

Simple calculators only need a single chip. The magnified picture on the right shows it has different circuits for each of the jobs it has to do.

How a calculator works out sums

Did you know that a calculator does all its calculations just by adding? This is the only mathematical process it can carry out in binary. It does multiplications, square roots and every other calculation by adding the numbers according to sets of rules stored in its permanent memory. Below you can see how a calculator works out 6 × 5.

X REGISTER

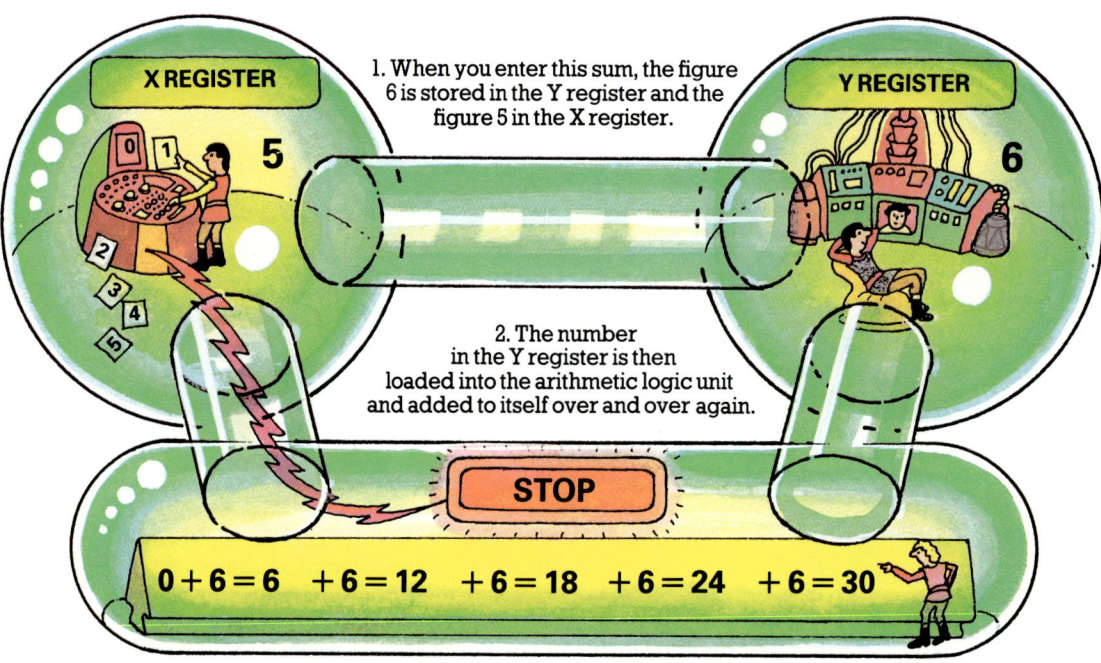

Y REGISTER

1. When you enter this sum, the figure 6 is stored in the Y register and the figure 5 in the X register.

2. The number in the Y register is then loaded into the arithmetic logic unit and added to itself over and over again.

STOP

0 + 6 = 6   + 6 = 12   + 6 = 18   + 6 = 24   + 6 = 30

3. The number in the X register shows how many times the addition should be done. Each time 6 is added to itself the X register number is reduced by 1.

4. When the number in the X register reaches zero a message tells the arithmetic logic unit to stop adding and the calculation is complete.

Every calculation the calculator does has to be reduced to simple steps which involve only adding. The rules for each calculation are called algorithms and they are worked out by the engineers and mathematicians who design calculators. The rules for multiplication are simple compared with those for square roots and other scientific functions, which may have hundreds of steps.

Calculator versus brain race

A calculator works so quickly that you do not notice how many steps it has to perform to do a sum. However, if you are good at mental arithmetic you may be able to beat the calculator. Ask someone to work out the sums on the right on a calculator while you do them in your head. Both start at the same moment. Can you do them faster than the calculator?

4 × 12

9 + 46

175 − 50

70 × 5

108 ÷ 12

4 × 7 + 5

9 + 8 − 3

73 − 17 + 16

244 ÷ 2 + 15

2 × 2 + 46

You can check your answers on page 93.

Number tricks

Here are some tricks to do on a calculator.

13 trick

This calculation gives the answer 13 whatever number you start with.

Think of any three-digit number, for example, 853

Enter it into the calculator	8 5 3
Then enter it again	8 5 3
Divide by 7	÷ 7
Divide by the original number	÷ 8 5 3
Divide by 11 and press =.	÷ 1 1
Try it with lots of other three-digit numbers.	= 13

Guess the number

Ask a friend to think of any number and write it down, without telling you what it is. Then give them the calculator and tell them to do the following:*

Enter the number (for example, 53)	5 3
Double it	× 2
Add 4	+ 4
Divide by 2	÷ 2
Add 7	+ 7
Multiply by 8	× 8
Subtract 12	− 1 2
Divide by 4	÷ 4
Subtract 11	− 1 1 =

Now take the calculator. Subtract 4 from the number in the display and divide by 2. The answer will be the number your friend first thought of.

− 4
÷ 2
= 53

Magic 9 trick

The answer to this calculation is 9 no matter what number you start with.

Write down any four digit number, for example, 5 279. Then jumble up the digits to make a new number, say 9 725.

Enter the larger of the two numbers into the calculator.   9 7 2 5 −

Subtract the smaller one from it and write down the answer.   5 2 7 9 = 4446

Then clear the calculator and add together the digits in the answer.

C 4 + 4 + 4 + 6 = 18

If the answer has more than one digit, clear the calculator again and add these together.

C 1 + 8 = 9

Try the trick with other four-digit numbers. The final answer will always be 9.

Making calculator words

If you turn a calculator upside-down the numbers look like letters. For example, you can write the word sizzle by entering 372215, then reading the numbers upside-down. Try 710, 77345, 3507 and 5376606 too. What words can you make?

*If you have a scientific calculator you may have to press = after each part of this calculation.

Using the memory

When you are doing calculations you often need to remember the answer to part of a calculation so you can use it again later on. Most calculators have a memory where you can store a number during a calculation. It is usually operated by three keys, labelled M+, M−, MR. Two other keys are described below. If the memory keys on your calculator are labelled differently check the instruction booklet to see what each one does.

The memory is an extra number store attached to the X register (see the picture on pages 58-59). It contains only zeros until you store a number in it. On most simple calculators the memory is erased when you switch off the calculator.

M+ / **MS** This key puts the number in the display into the memory. If there is a number already in the memory, the new number is added to it. Some calculators have an MS or "memory store" key. This puts the number in the display into the memory and erases any number already there.

M− This subtracts the number in the display from the number in the memory.

MR This is the memory recall key. It puts the number in the memory into the display so you can use it in the sum you are doing. The number is still stored in the memory though, so you can use it again if necessary.

MC The memory clear key erases a number from the memory. Some calculators have no memory clear key and to clear the memory you press MR followed by M−, or on some calculators, the ordinary clear key, followed by MS.

1 How to use the memory keys

[8] [4] [M+] *84*

To store a number, for instance, 84, in the memory first enter it into the calculator, then press M+ or MS. If you use the M+ key to store a number make sure the memory is clear before you start.

2

[C] [MR] *84*

Press C to clear the display. Then check the number is still in the memory by pressing MR. The number appears in the display and it is also retained in the memory.

3

[C] [6] [M+]
[MR] *90*

To add a number to the one in the memory enter it and then press M+. Press MR to see the new total.

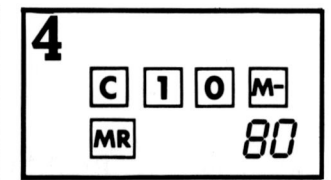

4

[C] [1] [0] [M−]
[MR] *80*

To subtract a number from the one in the memory enter it and then press M−. To check the new total in the memory press MR.

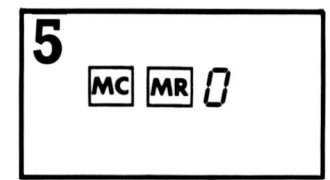

5

[MC] [MR] *0*

Clear the memory. Now if you press the MR key, a zero will appear in the display.

Memory sums

These examples show how you can use the memory to make calculations with several parts easier to solve. You can only store one number at a time in the memory so plan your calculations carefully. In divisions or subtractions you need to work out the second part of the sum first and store the answer in the memory so that you are dividing into or subtracting from the right part of the sum. There is an example like this in box 1. Before you start check that the calculator and the memory are clear.

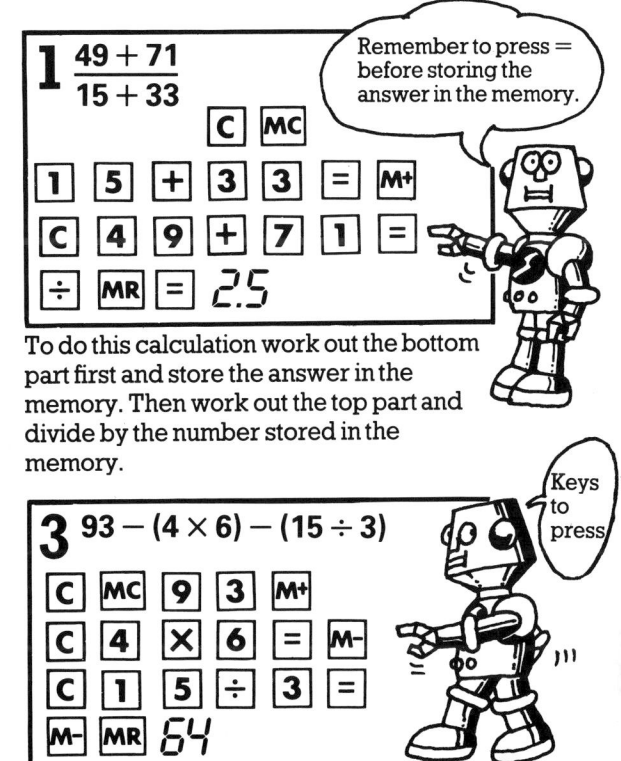

1 $\frac{49 + 71}{15 + 33}$

C MC
1 5 + 3 3 = M+
C 4 9 + 7 1 =
÷ MR = *2.5*

Remember to press = before storing the answer in the memory.

To do this calculation work out the bottom part first and store the answer in the memory. Then work out the top part and divide by the number stored in the memory.

2 16x table

Keys to press

C MC
1 6 M+ C
1 × MR =
2 × MR =

You can store a number in the memory to save entering it over and over again. For instance, to work out the 16× table store 16 in the memory, then press MR each time you multiply.

3 93 − (4 × 6) − (15 ÷ 3)

Keys to press

C MC 9 3 M+
C 4 × 6 = M−
C 1 5 ÷ 3 =
M− MR *64*

One way to do this calculation is to put 93 in the memory, then work out each of the sums in brackets and subtract the answers from 93 using the M− key. To get the final answer press MR.

Mountaineer puzzle

This mountaineer must not carry more than 17kg on his back. His rucksack with a tent, sleeping bag and climbing equipment weighs 11.75kg. To this he adds: cooking equipment weighing 2kg, two packets of rice weighing 430g each, seven packets of dried soup weighing 85g each, three packets of tea weighing 113g each, five sachets of milk powder weighing 21g each.

Then he remembers he should take some chocolate. Each bar weighs 103g. How many can he carry without exceeding his weight limit? Remember 1kg = 1 000g. The answer is on page 93.

Another kind of memory

Many simple calculators have another kind of memory called the constant function. This memory is automatic and there is no key to operate it. To check whether your calculator has a constant enter 5 + 2, then press =. The answer, 7, appears. Now press the = key again. If the answer changes to 9, the +2 part of the sum has been stored in the constant. It will be added to the figure in the display each time you press =. The constant is useful in sums where you need to perform the same function over and over again.

Below you can find out how to use the constant in calculations. Then you can see how to use it for doing conversions and for count-downs and keeping a tally.

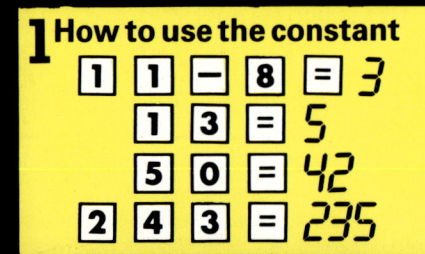

1 How to use the constant

1 1 − 8 = 3
1 3 = 5
5 0 = 42
2 4 3 = 235

Suppose you want to subtract 8 from 11, 13, 50 and 243. When you enter the first numbers, the −8 is stored in the constant. To subtract 8 from the other numbers, enter each number, then press =.

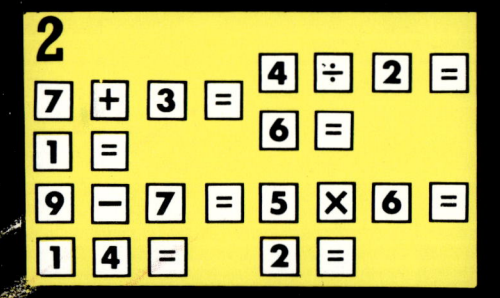

2

7 + 3 =   4 ÷ 2 =
1 =   6 =
9 − 7 =   5 × 6 =
1 4 =   2 =

Most calculators store the second part of an addition, subtraction or division in the constant, but the first part of a multiplication. Try the above examples to see how your calculator works.

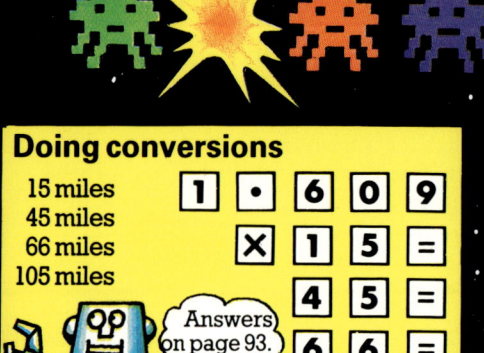

Doing conversions

15 miles
45 miles
66 miles
105 miles

1 . 6 0 9
× 1 5 =
    4 5 =
    6 6 =
1 0 5 =

Answers on page 93.

One mile is 1.609km, so to convert these distances to kilometres you multiply them by 1.609. When you do the first sum 1.609 is stored as a constant and to multiply the other numbers you enter them, then press =.*

Count-downs

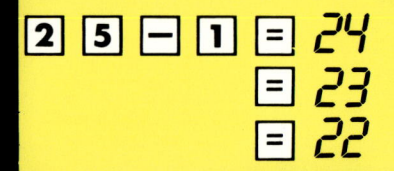

2 5 − 1 = 24
        = 23
        = 22

You can make a count-down by storing −1 in the constant. For example to count down 25 seconds to the start of a race enter 25 − 1, then press = each time a second passes.

Making a tally

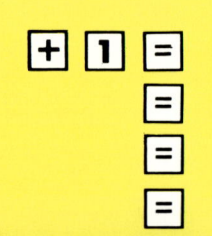

+ 1 =
  =
  =
  =

A tally is the opposite of a count-down. You could make a tally of the number of cars passing your house by entering +1, then pressing = each time a car passes.

*If your calculator stores the second part of a multiplication in the constant, you need to enter the first sum the other way round.

Space invader puzzle

Can you solve this puzzle? If your calculator has memory keys, try to do it without writing down any part of the calculations.

You are trying to beat a friend at space invaders. Their score is 18 950. You manage to knock out 151 invaders, 19 rockets, 5 flying saucers and the command ship. The scoring system is: 20 points for an invader, 330 points for a rocket, 550 points for a flying saucer and 5 000 points for the command ship.

Have you beaten your friend? What is the difference between your scores? (Answers on page 93.)

Change sign key

Keys to press

$$50 - \frac{4 \times 6}{2}$$

| 4 | × | 6 |
| ÷ | 2 | = | 12
| +/− | + | 5 | 0 | = | 38

If your calculator does not have memory keys you can use the change sign key (the key which changes positive numbers to negative ones and vice versa) to do sums such as the one shown here.

Work out the second part of the sum first. Then you have to subtract the answer from 50. To do this press the change sign key to make the number in the display negative. Now add 50. The calculation −12 + 50 gives the same answer as 50 − 12.

Error alert

Most calculators have a warning device to tell you if something is wrong. For instance, it is impossible to divide a number by 0. If you try to do so on a calculator an E standing for error, or a zero (sometimes flashing) appears in the display.

On most calculators the warning also appears if the answer to a calculation is too big or too small to be shown in the display. For example, if you try to divide 0.000 006 by 1 000 or multiply 5 000 000 by 20 000.

Working out percentages

If your calculator has a percent key (marked %) you can do percentages very easily and also work out percentage increases and reductions.

A percentage is a special kind of fraction. For example, 3% of 40 is the same as 3/100 of 40 and 25% of 70 is 25/100 (=1/4) of 70. The keys to press to work these out on a calculator are shown on the right.*

How to use the % key

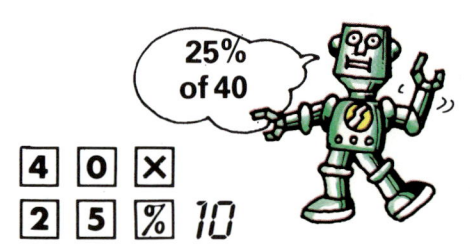

25% of 40

[4] [0] [×]
[2] [5] [%] 10

To work out a percentage of a number, you enter the number, then multiply by the percentage you want and press the percent key.

% increases and % reductions

Below you can find out how to work out percentage increases and reductions on a calculator. These calculations are useful for working out, say, the price of an item that has been reduced by 20% or how to add 10% service charge to a bill. You do these sums in different ways on different calculators. To find out which way you should use try these examples and check your answers with those given here.

1 [1][5][+]   or   [1][5][×]
　 [1][0][%]   [1][0][%]
　 [=] ← Some calculators   [+]
　 do not need =.

Here are the two ways to work out a 10% increase on 15. The answer should be 16.5.

2 [8][0][−]   or   [8][0][×]
　 [7][5][%]   [7][5][%]
　 [=] ← Some calculators   [−]
　 do not need =.

These are the two different sets of keys to press to reduce 80 by 75%. The answer should be 20.

Converting fractions to percentages

Keys to press   Answers
[3][÷][4][%]   75 %
[2][÷][3][%]   66.7 %
[5][÷][8][%]   62.5 %

On some calculators you can use the % key to change fractions to percentages. The keys to press to change 3/4 to a percentage are shown above. Try working out 2/3 and 5/8 too to see how it works on your calculator.

Hats puzzle

In the picture below can you work out what percentage of the robots have blue hats, what percentage have red hats and what percentage have yellow hats? Round off the percentages to whole numbers. (Answers on page 94.)

If you add up the three percentages the answer should be 100%. In fact it is only 99% because rounding off makes the percentages slightly inaccurate.

*For more about percentages and fractions, see pages 24-25.

Galactic spore puzzle

Planet Zanof has been invaded by a deadly weed – a spore from galactic space. The Zanof scientists are desperately trying to find a herbicide to destroy it, but the weed has already covered 1/5 of the planet, and every day the area it covers increases by 33%. How many days do the scientists have left before it overruns the whole planet?

Hints for working it out

The planet will be completely overrun when the weed covers 100% of it. At present the weed covers 20% (= 1/5), so enter 20 into the calculator and add 33% to it. Keep increasing the number in the display by 33% until it reaches just less than 100. The number of times you add 33% is the number of days they have left. You can check your answers on page 94.

Percentages without a % key

Here is an easy way to work out percentages if your calculator does not have a % key. Below you can find out how to work out 20% of 16.

$$20\% = \frac{20}{100} = 0.2$$

$$16 \times 0.2 = 3.2$$

To work out this sum first you have to convert the percentage to a decimal figure by dividing it by 100. An easy way to do this is to move the decimal point two places to the left, i.e. 20% is 0.2.

Then to work out 20% of 16 multiply 16 by the decimal number as shown above. See if you can work out the following percentages without using the % key: 25% of 5800, 35% of 675, 115% of 50, 46% of 900. (Answers on page 94.)

69

Squares and square roots

On these two pages you can find out how to use the square and square root keys on a calculator. Squaring a number means multiplying it by itself, and the symbol for squaring is a small 2 beside the number (for example, 4^2). On a calculator the square key is labelled x^2.

A square root is the opposite of a square. The label on the square root key is $\sqrt{}$ or $\sqrt{x}$.

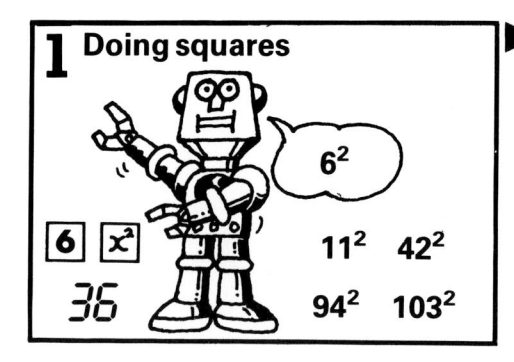

1 Doing squares

6^2

11^2 42^2

94^2 103^2

To square a number, for instance 6, enter it, then press the x^2 key. You should not press =. Can you work out the squares of the other numbers shown above? The answers are on page 94.

Doing square roots

$\sqrt{169}$

$\sqrt{15}$ $\sqrt{88}$ $\sqrt{529}$ $\sqrt{1944}$

To find the square root of a number you enter it, then press the $\sqrt{}$ key. Try the examples above. The answers are on page 94.

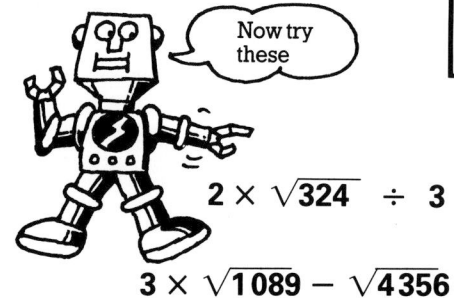

Now try these

$2 \times \sqrt{324} \div 3$

$3 \times \sqrt{1089} - \sqrt{4356}$

$9555 \div \sqrt{5402.25}$

In these calculations you can work out each square root as you come to it, but remember, always enter the number first, then press the $\sqrt{}$ key.

Square number tricks

Here are some tricks and puzzles to try. They all involve squaring numbers.

Reappearing squares puzzle

$5^2 = 25$ $6^2 = 36$ $25^2 = 625$

Above there are three numbers which when squared reappear as the last figure or figures in the answer.

There are only three other numbers between 25 and 1 000 which do this. Can you work out which they are? The answers are on page 94.
Hint: note that the numbers already given all end with a figure 5 or 6.

Time your reactions

How quickly can you catch a falling object? Here is a test to find out. You need a ruler, a calculator and a friend to help you.

1

Ask your friend to hold the ruler with the 0cm mark pointing down. Hold your thumb and finger round the ruler at the 0cm mark without actually touching the ruler. Then tell your friend to drop the ruler without warning you.

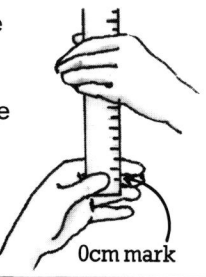

0cm mark

2 Can you do these?

0.2^2
0.65^2
0.83^2
0.9^2

If a number is less than 1, squaring it makes it smaller. Try the examples above. The answers are on page 94.

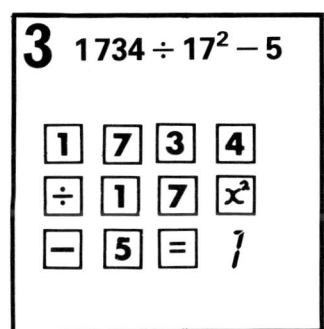

3 $1734 \div 17^2 - 5$

1	7	3	4
÷	1	7	x^2
−	5	=	1

This example shows how to work out squares in the middle of a calculation.

4 Keys to press

7^2

| 7 | × | = | 49 |

18^2

| 1 | 8 | × | = | 324 |

If you do not have an x^2 key you can work out squares by doing a normal multiplication, or you can use the constant* as shown here.

Back to square one

Here is a squaring sum to which the answer is always the number you started with.

Enter any number into the calculator, square it and store the result in the memory (or write it down).

Then enter the original number plus 1, and square that.

Subtract the square of the first number (which is in the memory).

Then subtract 1, and divide by 2.

Now try other numbers.

| 4 | 1 | x^2 |
| M+ |
| C | 4 | 2 |
| x^2 |
−	MR
−	1	÷
2	=	41

5^2 trick

Enter any number ending in 5. Square it, then store the result in the memory.

Now enter the number again without the figure 5 and multiply it by the next biggest number. Write down the answer.

| 6 | 5 |
| x^2 | M+ |

| C | 6 | × | 7 | = | 42 |

Then tack on 25 (which is 5^2). This number is the square of the number you first entered. Press MR to check it.

4225

| MR |

4225

Calculating your reaction time

2

Distance fallen

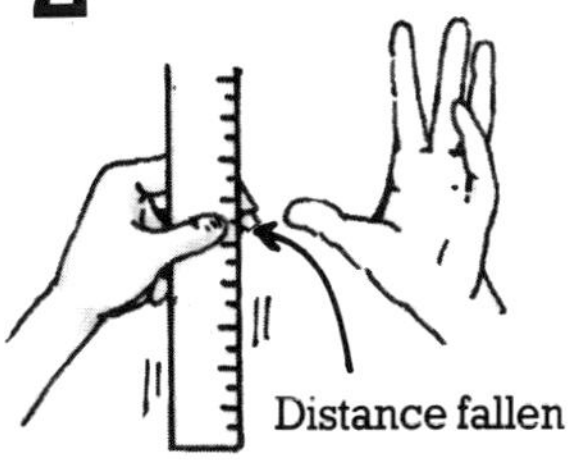

As soon as the ruler falls, catch it and note where your thumb and finger are gripping it. This shows the distance it has fallen.

Suppose the distance is 19.5cm. Your reaction time is the time it took the ruler to fall this distance. Since the ruler is falling under the pull of gravity its acceleration is 980cm/sec². So you can work out the time using the following mathematical formula:

$$\text{Time}^2 = 2\left(\frac{\text{distance}}{\text{acceleration}}\right)$$

Keys to press

| 1 | 9 | . | 5 | ÷ | 9 | 8 | 0 | × | 2 |
| = | 0.0397959 |

The answer is the time², so the time is $\sqrt{0.0397959}$, which is 0.2 seconds. This is quite a slow reaction time. Is yours better?

*See page 66. More square puzzles over the page. ▶

Square puzzles

The square and square root keys are useful for working out measurements of triangles using Pythagoras' theorem. Pythagoras' theorem states that on a right-angled triangle the square of the long side (called the hypotenuse) is equal to the sum of the squares of the other two sides. You can check this on the triangle below.*

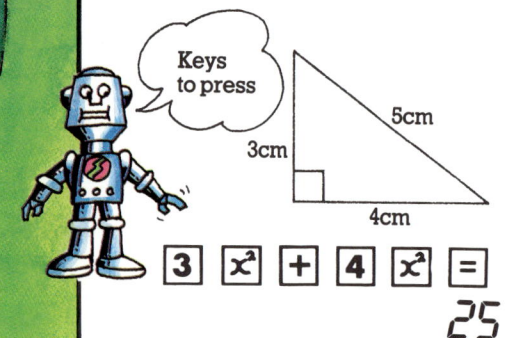

Keys to press

3cm, 4cm, 5cm

$3\ x^2\ +\ 4\ x^2\ =\ 25$

Now see if you can solve these puzzles. The answers are on page 94.

Slim Sally's CB aerial

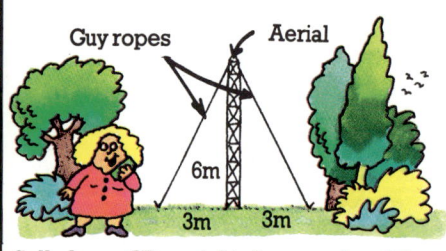

Guy ropes, Aerial, 6m, 3m, 3m

Sally has a CB aerial in her garden. The aerial is 6m high and her garden is 6m wide. How long are the guy ropes?

Walking puzzle

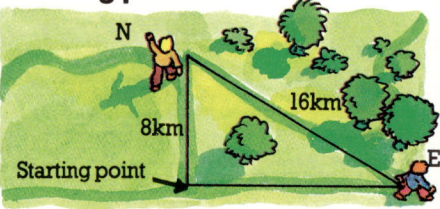

N, 8km, 16km, Starting point, E

Jim and Lee set off walking from the same point. Jim walks due North and Lee due East. When Jim has walked 8km he is 16km from Lee. How far has Lee walked?

*For more about Pythagoras' theorem, see page 27.

Star warrior puzzle

The evil Toron has captured an electric star warrior and imprisoned him in a water chamber. When the chamber is full the water will touch the warrior and electrocute him. The main pipe takes 9 minutes to fill the chamber, but the Toron sees a rescue ship coming so he turns on a second pipe as well. This takes 24 minutes to fill the chamber. The rescuers arrive after 6 minutes – are they in time?

Hints for working it out

One way to solve the puzzle is to work out how long it takes to fill the chamber using both pipes. Say it takes n minutes, then in one minute $1/n$ of the chamber is filled. You know that one pipe fills the chamber in 9 minutes, so in one minute it fills $1/9$ of the chamber. The other pipe fills $1/24$ of the chamber in one minute. So $1/n = 1/9 + 1/24$. To work out sums like this some calculators have a key which automatically divides numbers into 1. It is called the reciprocal key and labelled "$1/x$". You can use it to work out

More about reciprocals

A reciprocal is the mathematical name for the number you get when you divide a number into 1. For instance, the reciprocal of 2 is 0.5.

On some calculators the automatic constant divides numbers into 1 and you can use it to work out reciprocals. Press the keys shown above to see if it works on your calculator.*

Reciprocal trick

Try working out the reciprocals of 11 and multiples of 11 (for example, 22, 33, 55). They all contain two numbers which repeat themselves. One exception is 77.

Chamber gradually filling with water.

the value of $1/n$ like this:

[9] [1/x] [+] [2] [4] [1/x] [=]
0.1527778

If $1/n = 0.1527778$, then $n = \dfrac{1}{0.1527778}$

To work this out you can press the $1/x$ key again. This gives you n, the time it takes to fill the chamber. Did the rescuers arrive in time? You can check your answer on page 94.

If you do not have a reciprocal key you will need to work out the division into 1 each time and store the answer in the memory.

Rearranging sums

$$\frac{7}{1.1 + 2.5}$$

If your calculator does not have memory keys you can use the reciprocal key to work out sums like the one shown above in one operation. To do it you need to rearrange the sum like this:

$$\frac{7}{1.1 + 2.5} = 7 \times \frac{1}{1.1 + 2.5}$$

[1] [.] [1] [+] [2] [.] [5]
[=] [1/x] [×] [7] [=] 1.9

Then you can work out the bottom part of the fraction and press the reciprocal key and multiply by 7.

*If you get the answer 1, the automatic constant has divided 2 by itself, so you cannot use it to work out reciprocals.

Using the Pi key

Some calculators have a key labelled "π" (pronounced Pi). Pi is a never-ending number, discovered by the Ancient Greeks, which is used for working out the circumference and area of circles (see pages 16 and 122). Rounded off to seven decimal places it is 3.1415927.

Calculators with a π key have the number Pi stored in a special memory. When you want to use the number you can press the π key to put it into the display. If you have no π key and have to enter the number each time a shortened version, 3.14, is accurate enough for most calculations.

Working out circumferences

Circumference is distance round edge.

| 2 | × | π | × | 3 | = |

18.8 m

The mathematical formula for working out the circumference of a circle is $2\pi r$ (r stands for radius – the distance from the centre to the edge). Above are the keys to press to work out the circumference of a circle with a 3m radius.

Working out areas

Area

| π | × | 5 | x^2 | = |

78.5 m²

The formula for the area of a circle is πr^2. These are the keys to press to work out the area of a circle with a radius of 5m. If you do not have an x^2 key, enter the radius first, multiply it by itself and then multiply by π.

Walking the equator puzzle

How many steps would you take if you walked round the equator? Assume your steps are 0.5m long and use 6 370km as the radius of the Earth.

The answers are on page 94.

Bicycle puzzle

The radius is half the diameter.

The robot's bicycle wheel is 66cm in diameter. How many times must it turn to travel 1km? (1km = 100 000cm).

If the wheel turns 120 times each minute, what is the bike's speed in km/h?

Petrol thieves puzzle

A gang of petrol thieves is at work while the tanker driver is asleep. There are 50 petrol cans to be filled and only 8 minutes before the police catch up with them. Each can measures 0.5m in diameter and 1m in height. The petrol flows out of the tanker at a rate of 900 litres per minute. Can they fill all the cans and get away in time?
(Answer on page 94.)

Vital information

The formula for working out the volume of a petrol can is $\pi r^2 \times h$ (h is the height of the can). This gives you the volume in cubic metres. Multiply it by 1 000 to convert it to litres.

Spheres

You also need π to work out the surface area and volume of spheres.

Surface area

[4] [×] [π] [×] [2] [x^2] [=] 50.3m^2

The formula for working out the surface area of a sphere is $4\pi r^2$. Above you can see how to use it for a sphere with a radius of 2m.

Volume

[4] [÷] [3] [×] [π] [×] [6] [x^2] [×] [6] [=] 904.8m^3

The formula for the volume of a sphere is $\frac{4}{3}\pi r^3$. The small 3 above the r means "cubed", that is, multiplied by itself three times. Here are the keys to press to find the volume of a sphere with a 6m radius.

Bubble puzzle

If the circumference of this bubble is 63mm, what is its surface area? (Answer on page 94.)

How to use a scientific calculator

A scientific calculator has lots of extra keys for complex mathematical functions. It can carry out these functions automatically because the rules for doing them are stored in the circuits of its permanent memory.

The keys and labels on scientific calculators vary a great deal and yours may not look exactly like the ones shown here. You may need to check how the keys on your calculator work in your instruction booklet.

Enter key

Some calculators have no = key. These ▶ have an unusual system for entering numbers called Reverse Polish Notation (it was invented by a Polish mathematician). You enter all the numbers first, pressing a key marked "enter" after each number, then you press the appropriate function key.

Inverse key

The display often has space for 10 or more digits on scientific calculators.

◀ Double function keys

Many of the keys on a scientific calculator do more than one job. To make a key perform its second function you have to press a key marked "2ndF" or "INV" (short for inverse). The label inverse is used on some calculators because the second function is often the opposite of the first. For example, squaring a number is the opposite of finding a square root, so these two functions are on the same key.

What numbers are missing from this calculation $93 \times 8 = 7\_8$? (Answer on page 94.)

The symbols in red show the second function.

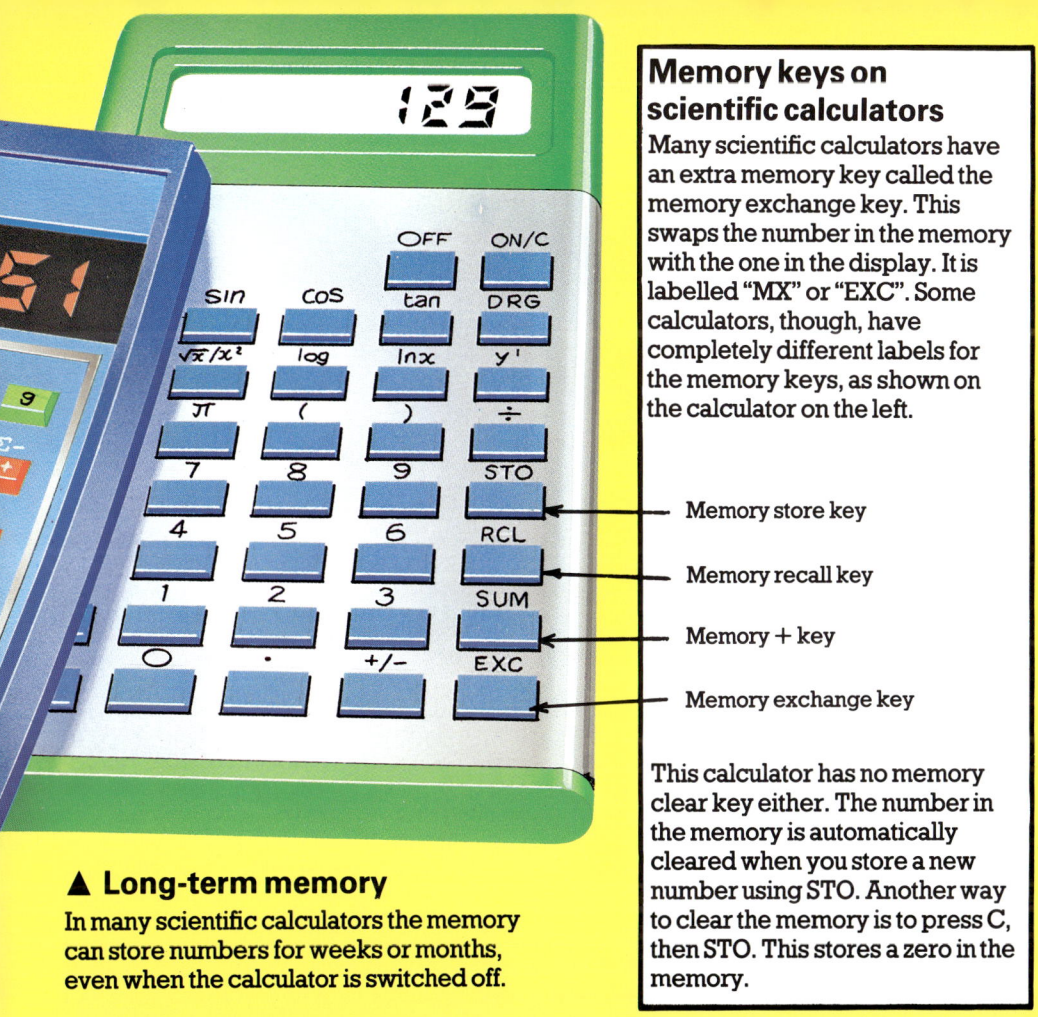

Memory keys on scientific calculators

Many scientific calculators have an extra memory key called the memory exchange key. This swaps the number in the memory with the one in the display. It is labelled "MX" or "EXC". Some calculators, though, have completely different labels for the memory keys, as shown on the calculator on the left.

— Memory store key

— Memory recall key

— Memory + key

— Memory exchange key

This calculator has no memory clear key either. The number in the memory is automatically cleared when you store a new number using STO. Another way to clear the memory is to press C, then STO. This stores a zero in the memory.

▲ Long-term memory

In many scientific calculators the memory can store numbers for weeks or months, even when the calculator is switched off.

Using the constant on a scientific calculator

On scientific calculators the constant is not automatic. To bring it into action there is a special key labelled "K".

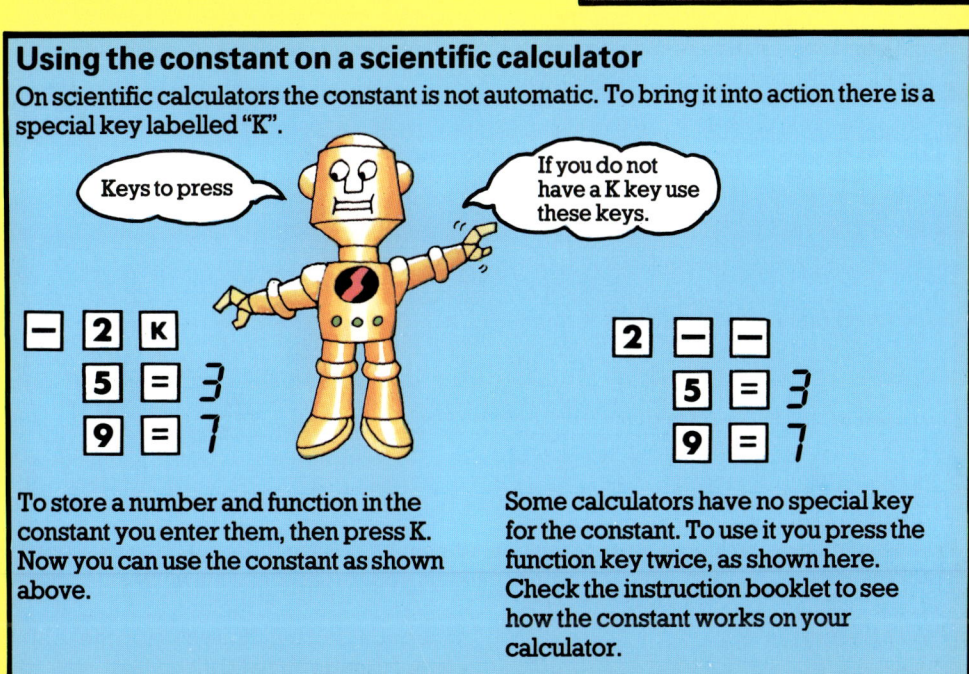

Keys to press

If you do not have a K key use these keys.

— 2 K
5 = 3
9 = 7

2 — —
5 = 3
9 = 7

To store a number and function in the constant you enter them, then press K. Now you can use the constant as shown above.

Some calculators have no special key for the constant. To use it you press the function key twice, as shown here. Check the instruction booklet to see how the constant works on your calculator.

77

Calculator logic

When calculations have several parts they can have more than one answer. For example, the answer to $4 + 3 \times 2$ could be 14 (if you add $4 + 3$ first) or 10 (if you multiply 3×2 first). To avoid confusion mathematicians have rules for sums with several parts. These are: do the division parts first, then the multiplications, then additions, then subtractions. This order of working is called algebraic logic.

Some calculators use algebraic logic and will always do divisions and multiplications before additions and subtractions regardless of the order in which you enter the sum. To check whether your calculator uses algebraic logic enter $4 + 3 \times 2 =$. If it does, it will give the answer 10.

To change the order of a calculation mathematicians use brackets to show which parts should be worked out together.

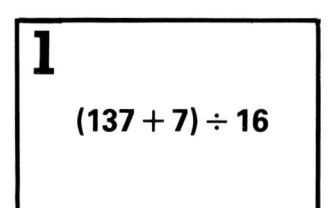

1

$(137 + 7) \div 16$

In this sum the brackets show that you should work out the addition part before dividing by 16.

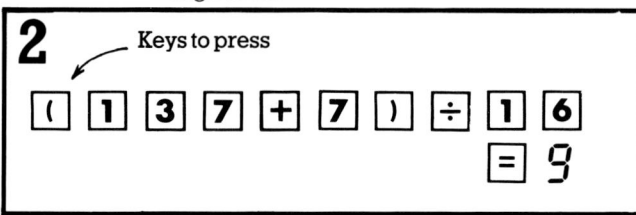

2

Keys to press

$($ 1 3 7 $+$ 7 $)$ $\div$ 1 6
$=$ 9

Scientific calculators have brackets keys, labelled (and), for doing calculations with brackets. You can see how to use them above. When you press the close bracket key the calculator works out the bit inside the brackets before going on.

Trying out the brackets keys

Here are some calculations you can do to try out the brackets keys. Where there are several sets of brackets in a calculation you should work out the functions between the sets of brackets in the correct mathematical order. A calculator with algebraic logic does this automatically so you can enter the sum in the order it is written. If your calculator does not use algebraic logic you must rearrange the sum and enter the different parts in the correct mathematical order.

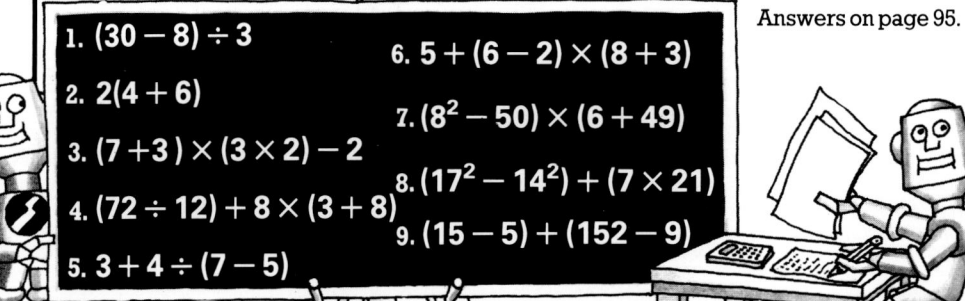

Answers on page 95.

1. $(30 - 8) \div 3$
2. $2(4 + 6)$
3. $(7 + 3) \times (3 \times 2) - 2$
4. $(72 \div 12) + 8 \times (3 + 8)$
5. $3 + 4 \div (7 - 5)$
6. $5 + (6 - 2) \times (8 + 3)$
7. $(8^2 - 50) \times (6 + 49)$
8. $(17^2 - 14^2) + (7 \times 21)$
9. $(15 - 5) + (152 - 9)$

When a number is written in front of the brackets with no function, as in $2(4 + 6)$ it means $2 \times (4 + 6)$ and you must press "×" when entering the sum.

If your calculator does not have brackets keys you can do these sums by entering the part inside the brackets first and, if necessary, storing the answer in the memory until you need it.

Nested brackets

$9 - 28 \div (70 \div (4 \times 2 + (3 - 1)))$

The calculation shown above has brackets within brackets. These are called nested brackets. If your calculator has brackets keys you can enter sums with nested brackets as shown below. The calculator automatically works them out in the correct mathematical order, which is to start with the innermost brackets and work outwards. You should work out the functions outside the brackets in the order of algebraic logic.

Most scientific calculators can cope with at least six sets of nested brackets. On some calculators a brackets symbol and a number are displayed each time you press the open brackets key, to show which set of brackets you are on.

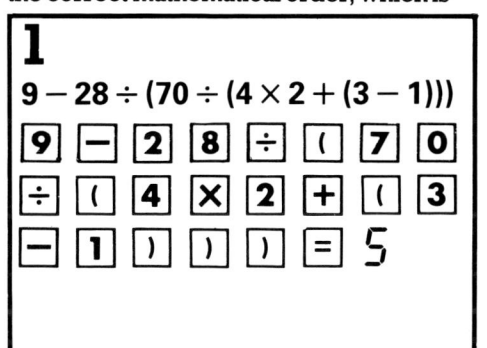

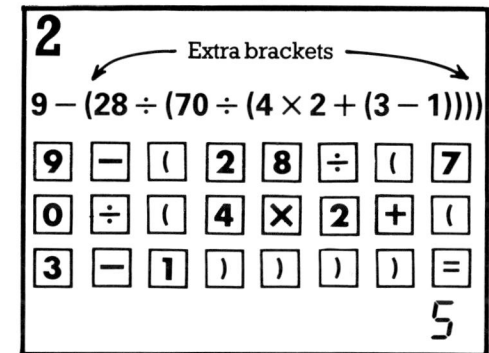

1
$9 - 28 \div (70 \div (4 \times 2 + (3 - 1)))$

[9] [−] [2] [8] [÷] [(] [7] [0]
[÷] [(] [4] [×] [2] [+] [(] [3]
[−] [1] [)] [)] [)] [=] 5

To do this calculation on a calculator with algebraic logic you enter it exactly as it is written and the calculator will work it out in the correct mathematical order.

2
Extra brackets
$9 - (28 \div (70 \div (4 \times 2 + (3 - 1))))$

[9] [−] [(] [2] [8] [÷] [(] [7]
[0] [÷] [(] [4] [×] [2] [+] [(]
[3] [−] [1] [)] [)] [)] [)] [=]
5

If your calculator does not use algebraic logic you must put in an extra set of brackets as shown above. This makes the calculator do the first division before the subtraction outside the brackets.

More calculations with brackets

Here are some examples for you to practise using the nested brackets on your calculator. If your calculator does not have algebraic logic, remember to insert extra brackets where necessary so the calculations are worked out in the correct mathematical order. (Answers on page 95.)

1. $5 + 6(7 - (2 \times 3))$
2. $73(40^2 - (23 \times (64 \div 4))) + 7 \times (50 - (5 \times 8))$
3. $(9 + 4) \div 3(7 \div (2 + 8))$

A calculator with a game

This calculator has a built-in "speed-shoot" game, as well as the usual mathematical functions. To play it, you switch the calculator into a special mode using a key marked "game".

Numbers start to move across the display from the right. The aim of the game is to knock out each number before it reaches the left-hand side and destroys one of your defences. You have an "aim" key which you have to press nine times to aim for a 9, three times for a 3 and so on, and a "fire" key. When all three defences are destroyed the game is over.

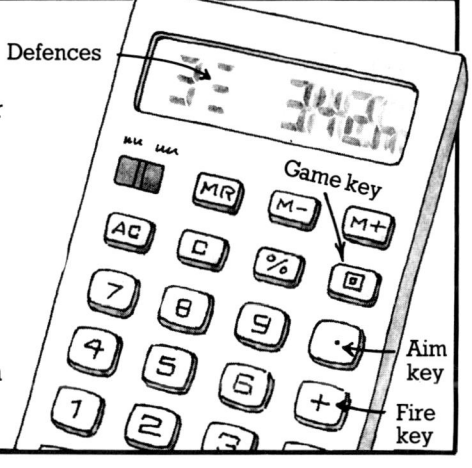

Angles and triangles

The keys labelled "tan", "sin" and "cos" (short for tangent, sine and cosine) are for working out measurements on right-angled triangles. They enable you to work out the lengths of the sides if you only know one angle and a side. To use these keys you enter the size of an angle, say 50°, then press the tan, cos or sin key.

[5] [0] [tan] 1.1917536
[5] [0] [sin] 0.7660444
[5] [0] [cos] 0.6427876

On some calculators it takes a moment for the answer to appear because the calculator needs "thinking time".*

To work out the measurements of a right-angled triangle there are some mathematical formulae using the tan, cos or sin of an angle.

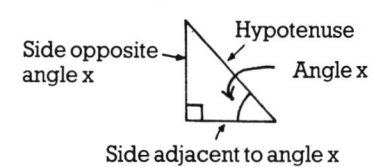

1. $\dfrac{\text{Opposite side}}{\text{Adjacent side}} = \tan x$

2. $\dfrac{\text{Opposite side}}{\text{Hypotenuse}} = \sin x$

3. $\dfrac{\text{Adjacent side}}{\text{Hypotenuse}} = \cos x$

You can find out how to use these formulae below.

1 Working out measurements on triangles

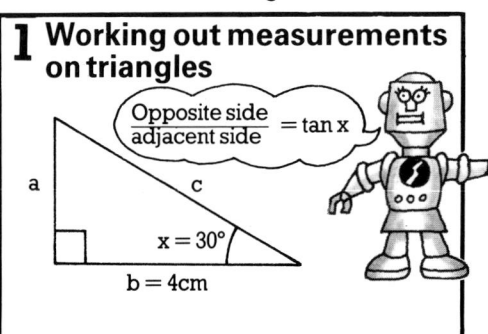

$\dfrac{\text{Opposite side}}{\text{adjacent side}} = \tan x$

On this triangle angle x is 30° and its adjacent side, b, is 4cm. To work out the length of side a (opposite side) you can use formula 1.

2

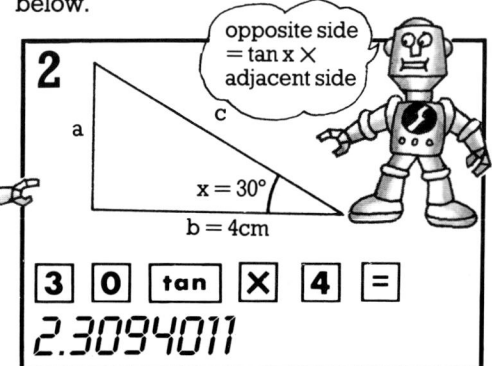

opposite side = tan x × adjacent side

x = 30°, b = 4cm

[3] [0] [tan] [×] [4] [=] 2.3094011

You know two of the measurements in the formula, so by turning the equation round you can work out that a (opposite side) = tan 30° × 4 (adjacent side).

3

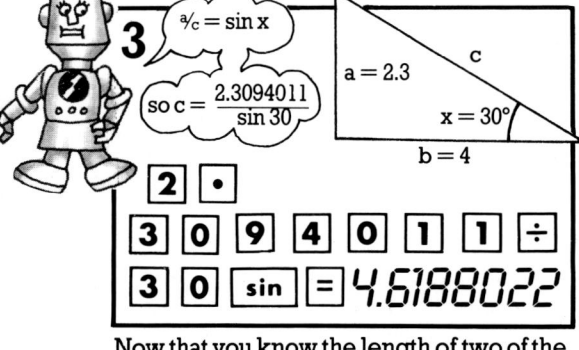

$a/c = \sin x$

so $c = \dfrac{2.3094011}{\sin 30}$

a = 2.3, x = 30°, b = 4

[3] [0] [9] [4] [0] [1] [1] [÷]
[3] [0] [sin] [=] 4.6188022

Now that you know the length of two of the sides you can work out the hypotenuse using formula 2 or 3. Do not round off the answer for side a if you use it in this calculation. The keys to press for formula 2 are shown above.

Check your calculations

c = 4.6, a = 2.3, b = 4, x = 30°

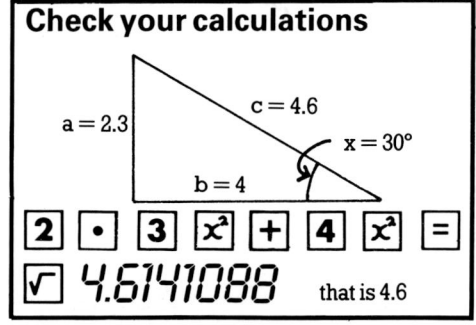

[2] [.] [3] [x^2] [+] [4] [x^2] [=]
[√] 4.6141088 that is 4.6

Now that you have worked out all the sides of the triangle you can check your calculations using Pythagoras' theorem (see pages 27 and 72). According to this, $a^2 + b^2 = c^2$. The keys to press for this calculation are shown above.

80

*If your answers are different from those above read the note on "Working in degrees" on the opposite page.

Yachtsman puzzle

A yachtsman on a stormy sea wants to find out how far he is from a dangerous headland. He works out that, from where he is, the beam of light from the lighthouse makes an angle of 40° with the sea level. The lighthouse is 45m high and is standing on a cliff 30m above sea level. How far is he from the headland? The answer is on page 95.

Hint

The top of the lighthouse, the yacht and the headland make the points of a right-angled triangle.

Working in degrees

Angles are usually measured in degrees (° is the symbol for degrees), but they can be measured in two other sets of units, just as length can be measured in metres or yards. The other units are called radians and gradients. One radian is 57.3° and one gradient is 0.9°.

On most scientific calculators there is a key or switch labelled "D.R.G", which sets the calculator to work in whichever unit you want. Each time you press the D.R.G key the unit changes and a symbol or letters (usually DEG, RAD or GRA) appear to show which units are being used. Make sure the calculator is using the correct units before you start doing calculations with angles.

Degrees can be subdivided into minutes and seconds. Some calculators have an extra key for entering degrees, minutes and seconds.

Working backwards

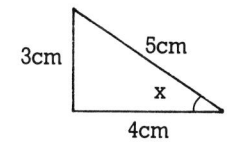

Sometimes you know the lengths of the sides of a triangle, and want to work out its angles. You can do this using the mathematical formulae given previously. On this triangle the sides measure 3cm, 4cm and 5cm. You can find angle x using any of the formulae. If you use formula 1,

$$\frac{\text{opposite side (3)}}{\text{adjacent side (4)}} = \tan x$$

So $\tan x = 0.75$.

Your answer is the tan of x. You can convert this to an angle by pressing the inverse key* and then the tan key. The keys to press are shown here.

[0] [.] [7] [5] [INV] [tan]

$$36.869898$$

You could also find x by working out its sine or cosine. $\sin x = 3/5 = 0.6$ and $\cos x = 4/5 = 0.8$, so to work out x the keys to press are:

[0] [.] [6] [INV] [sin]
[0] [.] [8] [INV] [cos]

*Some calculators have a key labelled "ARC" which you use with the sin, cos and tan keys instead of the inverse key.

Working out powers and roots

Multiplying a number by itself over and over again is called raising it to a power. For example, if you multiply 4 by itself five times, that is, $4 \times 4 \times 4 \times 4 \times 4$ you are raising it to the power 5 (see page 38). The power shows how many times the number appears in the multiplication. It is written 4^5.

Raising a number to the power 2 is the same as squaring it. Raising it to the power 3 is called cubing it.

Most scientific calculators have a key for working out powers, labelled y^x (y is the number you want to multiply and x is the power).*

1 4^5

| 4 | y^x | 5 | = | 1024 |

Above you can see the keys to press to work out 4^5. Try out the power key on these numbers too: 9^8, 15^3, 6^1, 12^4, 5^7. The answers are on page 95.

2 $9^3 + 17 - 4^4$

| 9 | y^x | 3 | + | 1 | 7 |
| - | 4 | y^x | 4 | = | 490 |

You can usually work out powers in sums as you go along. Remember, the $+$, $-$, $\times$ and $\div$ keys act as $=$ keys so you do not need to press $=$ until the end of the sum.

Negative powers

The number 4^{-3} is called a negative power. You can work out negative powers on a calculator using the change sign key.

| 4 | y^x | 3 | +/- | = | 0.015625 |

In fact 4^{-3} is the same as $1/4^3$. Another way to work it out is to do 4^3, then divide the answer into 1 by pressing the reciprocal key as shown below.

| 4 | y^x | 3 | = | 1/x | 0.015625 |

3^{-4} 5^{-2} 4^{-1} 6^{-5}

Try working out these negative powers. The answers are on page 95.

3 $((75 - 30) \div 5)^7$

| (| (| 7 | 5 | - | 3 | 0 |) |
| $\div$ | 5 |) | y^x | 7 | = |

4782969

To do this sum work out everything inside the brackets first, then raise the result to the power of 7. Above you can see how to do it using the brackets keys (see page 79).

Working out roots

Roots are the opposite of powers. For example, $4^5 = 1024$ and the 5th root of 1024 is 4. Roots can be written in two ways: $\sqrt[5]{1024}$ or $1024^{1/5}$.

Some calculators have a root key labelled $\sqrt[x]{y}$ or $y^{1/x}$, but on many the root function is on the same key as the power function. To work out a root on these calculators you press the inverse key before pressing the power key, as shown on the right.

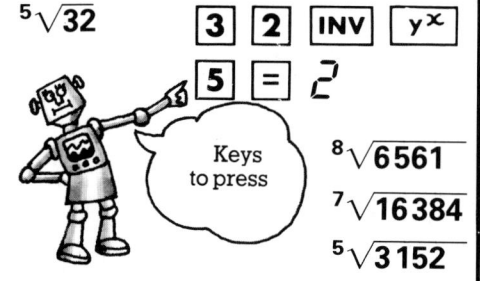

$\sqrt[5]{32}$

Keys to press

| 3 | 2 | INV | y^x |
| 5 | = | 2 |

$\sqrt[8]{6561}$

$\sqrt[7]{16384}$

$\sqrt[5]{3152}$

Above you can see the keys to press to work out the 5th root of 32. Can you work out the other roots? The answers are on page 95.

82

*On some calculators the label is x^y – so x is the number and y the power.

Taking risks*

Is it worth taking a risk? There is a mathematical way to work out how big a risk you are taking and what your chances are of winning. Say someone bets that they can cut a newly shuffled pack of cards and get the ace of spades. You can work out their chances of winning as follows. There are 52 cards in the pack and only one of them is the ace of spades, so their chances of winning are 1 in 52. You can write this as $1/52$ or as a decimal 0.019. If you convert the decimal to a fraction it is $19/1000$, so you should get the ace 19 times out of every thousand cuts.

Now try this one. If you toss a coin five times, what are your chances of getting five heads in a row.

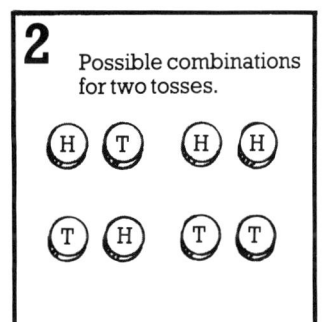

A coin has two sides so your chances of getting heads on the first toss are 1 in 2, or $1/2$.

2 Possible combinations for two tosses.

For two tosses there are four possible combinations of heads and tails. Only one of them is H, H, so your chances of getting two heads in a row are $1/4$.

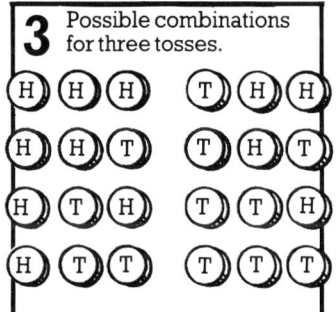

3 Possible combinations for three tosses.

When you toss the coin three times in a row there are eight possible combinations of heads and tails. Your chances of getting three heads are $1/8$.

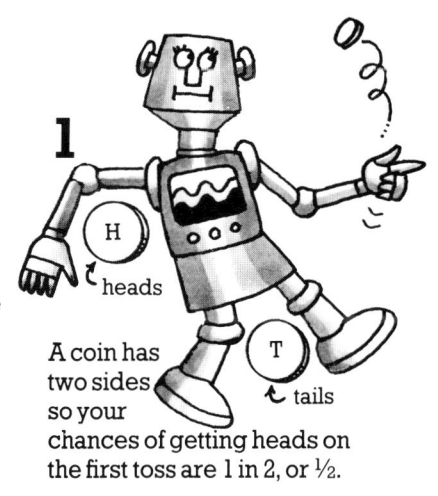

4 The fraction $1/4$ can also be written as $1/2^2$ and $1/8$ as $1/2^3$. With each extra toss of the coin the number of combinations of ways that the coin can fall increases by a power of 2. So the chances of getting five heads in a row are $1/2^5$. The keys to press to work this out are shown below.

$$1 \div 2 \; y^x \; 5 \; =$$

What are your chances? (Answer on page 95.)

Calculators in space

Space shuttle pilots take special calculators with them for doing complex calculations during the flights.

The calculators are programmable. In fact, they are like small computers with displays which show words as well as numbers.

The shuttle pilots take two calculators. One is programmed to work out which ground station the pilots can contact at every stage in the mission, and when and for how long the contact can be made. The second is for working out how the spacecraft should be balanced for re-entry into the Earth's atmosphere.

*You can find out more about this on pages 42-44 and 134-135.

Very large and very small numbers

Simple calculators cannot cope with numbers so large or so small that they have too many digits to fit into the display. However scientific calculators can do sums with these numbers using a kind of shorthand called scientific notation.

1 Scientific notation

$4\,000\,000 \times 221\,000$

If you do this multiplication, the calculator gives the answer as shown above. The answer is in scientific notation and it means 8.84×10^{11}. The 10^{11} part is called the exponent.

2 $8.84\ 11$

The exponent shows how many places you move the decimal point to change the answer to an ordinary number. In this case you move it 11 places to the right. (This is the same as multiplying 8.84 by 10^{11}.)

3 $1 \div 80\,000\,000$

To see what the calculator does with a very small number, try dividing 1 by 80 000 000. The calculator gives the answer in scientific notation and it means 1.25×10^{-8}.

4 $1.25\ -08$

When the exponent is negative you move the decimal point to the left. In this case you move the point eight places to the left as shown above. (This is the same as multiplying 1.25 by 10^{-8}.)

Sun puzzle

The Sun is 1.5×10^{11} metres away from Earth and the speed of light is 3×10^8 m/sec. How long does it take light from the Sun to reach Earth? Give the answer in seconds. (You can check your answers to these puzzles on page 95.)

Snail puzzle

If a snail moves at 0.0005km/h, how long would it take it to crawl to the Moon 384 000km away? Give your answer in years.

84

Doing sums in scientific notation

You can also use scientific notation to do calculations with numbers which are too large for the display. You have to enter the numbers in scientific notation and you can do this using an exponent key labelled "EXP" or "EE".

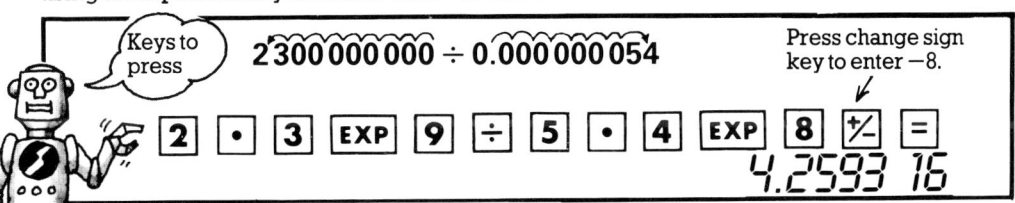

Keys to press: $2\,300\,000\,000 \div 0.000\,000\,054$

Press change sign key to enter -8.

`2` `.` `3` `EXP` `9` `÷` `5` `.` `4` `EXP` `8` `+/-` `=`

Display: 4.2593 16

To do the above calculation you first have to convert the numbers to scientific notation. To convert numbers above 1, you move the decimal point to the left until it comes after the first digit. To convert numbers below 1, you move the decimal point to the right until it comes after the first digit which is above 0. The number of times you move the decimal point is the exponent. If the number is below 1 the exponent is negative.

In the first number you move the point nine times to the left, so the scientific notation is 2.3×10^9. In the second number you move the point eight times to the right so the scientific notation is 5.4×10^{-8}. The keys to press to enter these numbers are shown above.

Examples to try
1. $650\,000\,000 \times 342\,000 \times 3\,098$
2. $1\,376\,000\,000 \div 0.000\,000\,08$
3. $8\,052\,000\,000 \times 23 \div 0.000\,000\,55$

Here are some more examples to try using scientific notation. If a number is small enough to fit into the display you can enter it in the normal way. Calculators which use scientific notation can cope with a mixture of scientific and ordinary numbers. (Answers on page 95.)

Unexpected answers
$(7.45 \times 10^{10}) + 23$

Answer: 7.45 10

See if you can do this example on a calculator. In its answer the calculator appears not to have added 23. This is because when you are dealing with such a large number 23 is too small a part of it to show up. In fact the answer is 74 500 000 023, but the nearest the calculator can get to it is 7.45×10^{10}.

Paper folding puzzle
Try to guess the answer to this puzzle before you work it out.

If you had a sheet of paper 0.15mm thick and folded it in half 50 times* how high would it be?

Hint
You have to double 0.15mm 50 times, which is the same as multiplying it by 2^{50}.

*In fact this is impossible. A square sheet of paper cannot be folded more than eight times – try it and see.

Simple statistics

Statistics are facts and figures which give information about groups of things. One way of giving information about groups is to work out an average. The term for the kind of average used most in maths is the "arithmetic mean".*

Some calculators have special keys for working out averages and other kinds of statistics. The keys are labelled differently on different calculators. The most common labels are shown here, but check the instruction booklet to find the statistics keys on your calculator.

Data entry key [x] or [Σ+]

Arithmetic mean key [x̄] [MEAN]

Standard deviation key [σn-1]

Statistics register clear key [SAC] [CSR]

These are the basic statistics keys, with the labels most frequently used. On some calculators you have to switch into a statistics register or "mode" before you can do statistics. When you are using the statistics keys other functions, such as the memory, may be out of action because the calculator needs them itself.

Using the statistics keys

To find out how to use the statistics keys try working out the average time of these six runners in a 100 metre race. The times are your "data" and they are shown in the picture below.

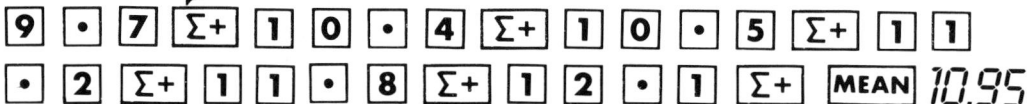

12.1secs
11.8secs
10.4secs
10.5secs
11.2secs
9.7secs

Data entry key

[9] [.] [7] [Σ+] [1] [0] [.] [4] [Σ+] [1] [0] [.] [5] [Σ+] [1] [1] [.] [2] [Σ+] [1] [1] [.] [8] [Σ+] [1] [2] [.] [1] [Σ+] [MEAN] *10.95*

Arithmetic mean key

To work out the average time, you add all the times, then divide by the number of runners. To do this on a calculator switch to the statistics register (if necessary) then press the statistics register clear key to make sure that there are no numbers left there from previous calculations. Enter the times of the runners using the data entry key. The keys to press are shown above. Then press the arithmetic mean key to make the calculator work out the average and show it in the display.

*You can find out more about this and other types of statistics on pages 30-31.

Entering data

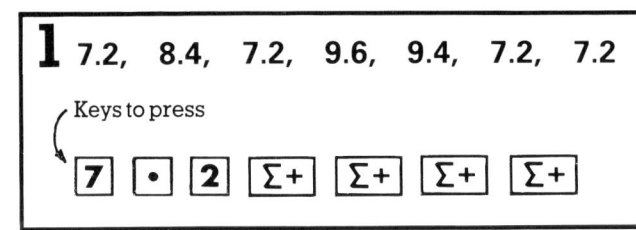

1 7.2, 8.4, 7.2, 9.6, 9.4, 7.2, 7.2

Keys to press

[7] [.] [2] [Σ+] [Σ+] [Σ+] [Σ+]

2

[3] [.] [8] [C]
[3] [.] [9] or [Σ−]

When you have lots of data some of the figures may be the same. In the example above you can group the figures so all the 7.2s can be entered together. To enter the figures, press the digit keys once, then press the data entry key the number of times the figure occurs.

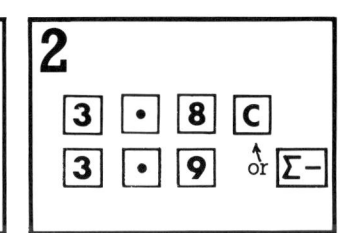

If you enter a number wrongly, you can clear it by pressing the usual clear entry key, or a special Σ− or "DEL" key.

More about averages

Averages may give misleading information if one or two of the figures on which they are based are much higher or lower than the rest. If the slowest runner took 23.4 instead of 12.1 seconds, the new average time would be 12.8 seconds. This is slower than five of the six runners, so it does not give a true picture of the speed of the race.

Mathematicians have a way of working out how much figures vary from the average. It is called finding the "standard deviation" of the average. The key for doing it on a calculator is labelled $\sigma n\text{-}1$.

23.4secs
11.8secs
11.2secs
10.4secs
10.5secs
9.7secs

Working out standard deviations

To work out a standard deviation, enter all the data then press the standard deviation key. Try it with the runners' times in the first example. (You should get 0.91secs.)

Now try the standard deviation for the second example. A higher figure for the standard deviation shows that the data is spread over a wider range.

Other statistics keys

Some calculators have several more statistics keys, such as the "sum of data squared" key labelled Σx^2 which is for more complex statistical calculations.

You may also have an Σx key which gives the sum of all the data entered and a key labelled "n" which you can press to check the number of data you entered. Some calculators automatically display a running total as you enter the data.

Permutations

In the picture on the right there are three robots and three seats. How many different arrangements of robots on seats do you think there are?

Different arrangements of things are called permutations. There is a mathematical way to work out how many permutations are possible. You can find out what it is below.

How to work it out

For the first seat there are three possible robots. For the second seat there are two possible robots in each case, and for the third seat there is one possible robot in each case. To work out the number of different seating arrangements you multiply the number of possible robots for each seat, that is $3 \times 2 \times 1$.

Working it out on a calculator

The calculation $3 \times 2 \times 1$ is a "factorial". It is called factorial 3 and is written 3!. Most scientific calculators have a key for working out factorials, labelled x! or n!. You can see how to use it above.

If there were seven robots and seven seats the number of possibilities for the first seat would be 7, for the second, 6, and so on. So the number of permutations would be 7!.

Choosing a space crew

Marlo wants a new crew for the spaceship. He needs three crew members and there are 17 to choose from. How many different crews could he have?

You need to find out how many combinations of three you can get from 17. Then you must take into account that some of these will be different permutations of the same three (ABC is the same as BAC, BCA, and so on). To work out how many *different* groups of three are possible, you have to divide the total number of combinations by the number of possible permutations of 3.

[1] [7] [×] [1] [6] [×] [1] [5]
[÷] [3] [x!] [=]

The total number of combinations is 17 × 16 × 15, because there are 17 possibilities for the first crew member, 16 possibilities for the second crew member and 15 for the third crew member. Divide this by the number of permutations of 3 that are possible, that is 3!, to get the answer. You should find that Marlo can have 680 different crews.

Picking teams

Imagine you have to pick two football teams of 11 people. There are 15 boys and 20 girls to choose from. How many different boys' and how many different girls' teams could you make? (Answers on page 95.)

Statistics puzzle

There are six statistics puzzles in this picture. Can you solve them? (When you work out an average give the standard deviation too.) The answers are on page 95.

1 What is the average number of people at a bus-stop?

2 If one bus queue waits three minutes for a bus, another waits 20 minutes and the third waits half an hour, what is the average time each queue waits for a bus?

3 What percentage of the people are waiting for a bus and what percentage are travelling by car?

4 How many different combinations of people from this queue could sit on the seat in the bus-stop? (The seat holds four people.)

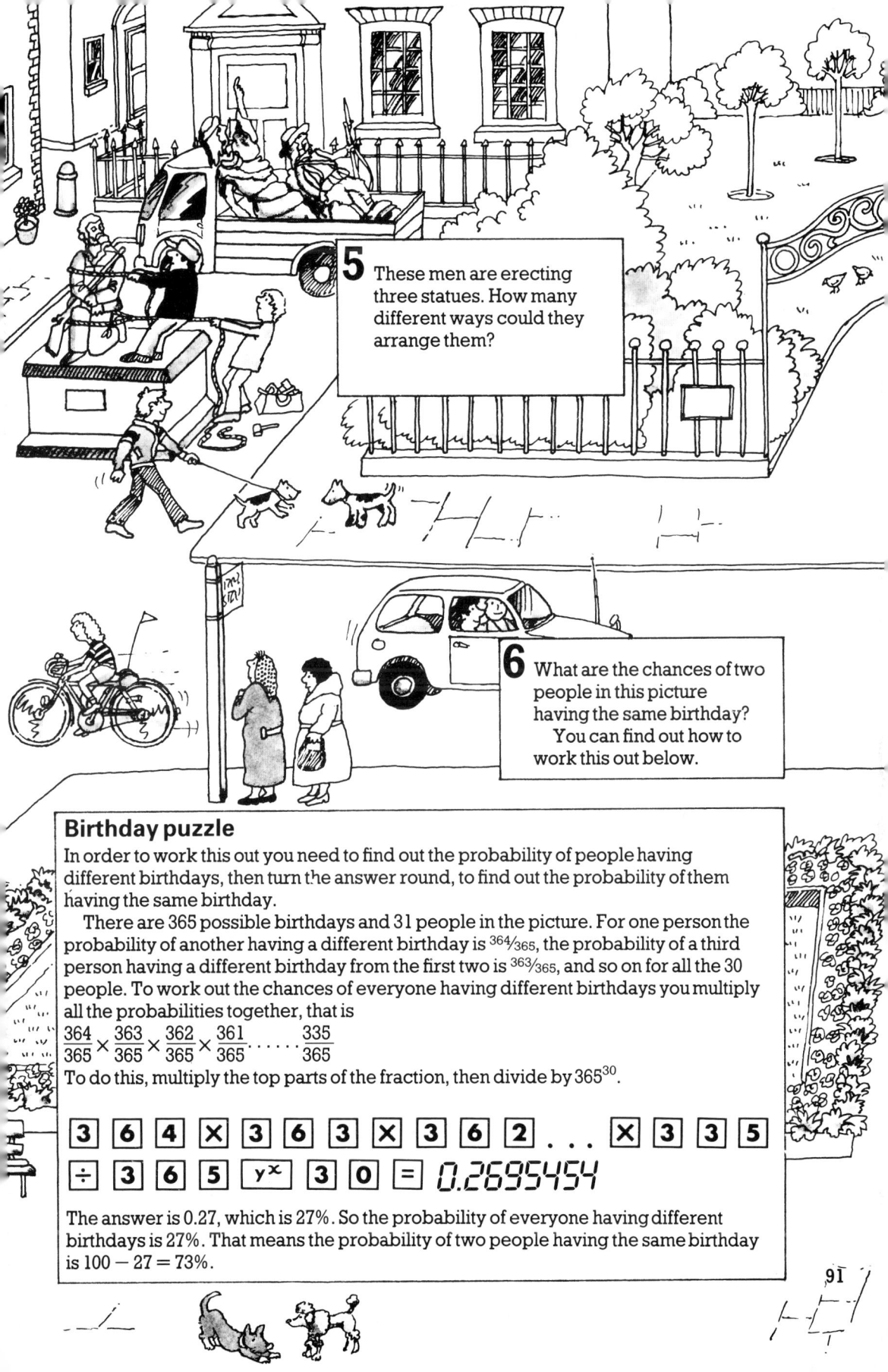

5 These men are erecting three statues. How many different ways could they arrange them?

6 What are the chances of two people in this picture having the same birthday? You can find out how to work this out below.

Birthday puzzle

In order to work this out you need to find out the probability of people having different birthdays, then turn the answer round, to find out the probability of them having the same birthday.

There are 365 possible birthdays and 31 people in the picture. For one person the probability of another having a different birthday is $364/365$, the probability of a third person having a different birthday from the first two is $363/365$, and so on for all the 30 people. To work out the chances of everyone having different birthdays you multiply all the probabilities together, that is

$$\frac{364}{365} \times \frac{363}{365} \times \frac{362}{365} \times \frac{361}{365} \cdots \cdots \frac{335}{365}$$

To do this, multiply the top parts of the fraction, then divide by 365^{30}.

$$\boxed{3\;6\;4} \times \boxed{3\;6\;3} \times \boxed{3\;6\;2} \ldots \times \boxed{3\;3\;5}$$
$$\div \boxed{3\;6\;5} \; y^x \; \boxed{3\;0} = 0.2695454$$

The answer is 0.27, which is 27%. So the probability of everyone having different birthdays is 27%. That means the probability of two people having the same birthday is $100 - 27 = 73\%$.

Other keys

Some scientific calculators have lots more keys not mentioned so far. Two of the most common are the logarithm, or log, keys labelled "log" and "ln". The log key gives the common logarithm of a number, that is, the number expressed as a power of 10. For example, 100 is 10^2, so log 100 is 2. The "ln" key gives the natural logarithm of a number. A natural logarithm is a number expressed as a power of this number: 2.7182818 (this number is called "e").

Logs were invented before calculators existed to make multiplying and dividing long numbers easier. Instead of multiplying you can add their logs and instead of dividing you can subtract their logs. Logs are still used in some scientific calculations.

A key marked a^b/c is for doing calculations with fractions without changing them to decimals.

Using the common log key

[1] [0] [0] [log] 2

To get the common log of a number you enter it, then press the log key as shown above.

[2] [INV] [log] 100

The log key usually has an "antilog" function on it too. This converts logs back to ordinary numbers. To use it, enter the log and press the inverse key, then the log key. The symbol for a common antilog is "10^x".

Using the natural log key

[1] [0] [0] [ln] 4.6051702

To find the natural log of a number you enter it, then press the ln key.

[4] [.] [6] [INV] [ln]
99.982983

The symbol for a natural antilog is "e^x". You can find natural antilogs using the inverse key as shown above. (The answer in this case may not be 100 because you entered a rounded off version of log 100.)

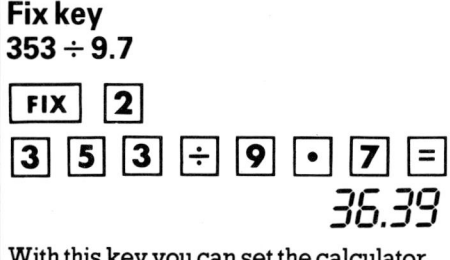

Fix key
$353 \div 9.7$

[FIX] [2]

[3] [5] [3] [÷] [9] [.] [7] [=]
36.39

With this key you can set the calculator to round off the answer to the number of decimal places you want. For example, if you want the answer to have two figures after the decimal point, press fix, then 2 before entering the calculation.

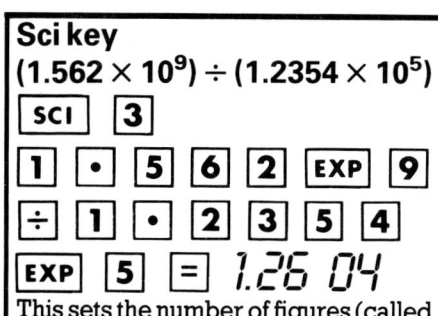

Sci key
$(1.562 \times 10^9) \div (1.2354 \times 10^5)$

[SCI] [3]

[1] [.] [5] [6] [2] [EXP] [9]
[÷] [1] [.] [2] [3] [5] [4]
[EXP] [5] [=] $1.26 \ 04$

This sets the number of figures (called significant figures) the calculator gives in the answer to a sum done in scientific notation. Most calculators can show up to five significant figures. Above you can see how to get an answer with three significant figures.

Puzzle answers

Most of the answers are rounded off to whole numbers or one decimal place. Where it is necessary to be more accurate the answer is given as it would appear on a calculator with an eight-digit display which rounds off the last digit.

If you have rounded off numbers before the end of a sum, your final answer may be slightly different from the ones given.

Page 50
The largest number you can produce is 22 412, that is 431 × 52.

Page 52
You can make the calculator display 50 using only 7, 5, +, − and = by doing the following calculation:
7 + 7 + 7 + 7 + 7 + 7 + 7 + 7 (= 56) − 5 (= 51) + 7 + 7 (= 65) − 5 − 5 − 5 = 50.
Or you can just enter 55 − 5. Can you think of any other ways?

Page 54
Simple sums
25 × 4 = 100; 1 100 × 9 + 9 = 9 909;
30 ÷ 2.4 − 3 = 9.5; 9 521 − 4 193 = 5 328;
10.4 ÷ 2 = 5.2; 6 400 ÷ 8 = 800;
6 × 10 × 100 = 6 000; 9 + 17 = 26;
54 ÷ 4.5 − 6 = 6; 700 × 4.3 ÷ 301 = 10;
805.2 − 605.2 = 200; 4 098 − 3 097 = 1 001;
6 059 × 423 = 2 562 957

Page 55
Rough guesses
512 × 359 = 183 808 (rough guess = 200 000)
971 × 28 = 27 188 (rough guess = 30 000)
1 594 + 273 = 1 867 (rough guess = 1 900)
6 123 ÷ 57 = 107.4 (rough guess = 6 000 ÷ 60)

Page 56
Car rally puzzle
Your average speed over the last section must be just over 98km/h. So it is unlikely that you can reach the check-point in time without breaking the speed limit.

The other car's average speed is almost 47km/h. That is the total distance (19.5km) divided by the time it takes (25 ÷ 60 hours).

Page 57
Converting fractions
$1/10$ = 0.1; $7\frac{3}{4}$ = 7.75; $3\frac{5}{8}$ = 3.625;
$9/8$ = 1.125; $5/17$ = 0.2941176;
$15/16$ = 0.9375; $21\frac{2}{3}$ = 21.666667

Page 61
Binary decoder
0011 = 3; 1111 = 15; 0101 = 5;
0111 = 7; 1100 = 12; 1001 = 9.

Page 62
Calculator versus brain race
4 × 12 = 48; 9 + 46 = 55; 175 − 50 = 125;
70 × 5 = 350; 108 ÷ 12 = 9; 4 × 7 + 5 = 33;
9 + 8 − 3 = 14; 73 − 17 + 16 = 72;
244 ÷ 2 + 15 = 137; 2 × 2 + 46 = 50

Page 65
Mountaineer puzzle
The mountaineer can take 13 bars of chocolate. One way to work this out is to calculate the weight (in kg) that he has already:
$$11.75 + 2 + (2 \times 0.43)$$
$$+ (7 \times 0.085)$$
$$+ (3 \times 0.113)$$
$$+ (5 \times 0.021)$$
Doing the additions in the memory will give you an MR total of 15.649. Subtract this from 17kg to find out how much more weight he can take, then divide by 0.103 to work out how many bars of chocolate that is.

Page 66
Conversions
  15 miles = 24.1km
  45 miles = 72.4km
  66 miles = 106.2km
  105 miles = 168.9km

Page 67
Space invader puzzle
You have not beaten your friend. The difference between your scores is 1,910 points. The best way to work it out is to calculate each part of your score and add them in the memory. Then you can press MR to find out your total score. To find the difference between the scores enter your friend's score and subtract yours from it by pressing "−", then MR.

Page 68
Hats puzzle
44% (4/9) of the robots have yellow hats, 33% (3/9) have red hats and 22% (2/9) have blue hats.

Page 69
Galactic spore puzzle
The scientists of Zanof have five days in which to find a herbicide. On the sixth day the planet will be completely overrun. Your calculations should show:

day 1   20 + 33% = 26.6% of planet covered
day 2   + 33% = 35.378%
day 3   + 33% = 47.05274%
day 4   + 33% = 65.5801442%
day 5   + 33% = 83.231592%
day 6   + 33% = 110.69802%

Percentages without a % key
25% of 5800 = 1450; 35% of 675 = 236.25; 115% of 50 = 57.5; 46% of 900 = 414.

Pages 70-71
Doing squares
$11^2 = 121$; $42^2 = 1764$; $94^2 = 8836$; $103^2 = 10609$.
$0.2^2 = 0.04$; $0.65^2 = 0.4225$; $0.83^2 = 0.6889$; $0.9^2 = 0.81$.

Reappearing squares
The three other numbers which, when squared, reappear as the last figures in the answer are $76^2 = 5776$; $376^2 = 141376$; $625^2 = 390625$.

Doing square roots
$\sqrt{15} = 3.9$; $\sqrt{88} = 9.4$; $\sqrt{529} = 23$; $\sqrt{1944} = 44.1$.

$2 \times \sqrt{324} \div 3 = 12$;
$3 \times \sqrt{1089} - \sqrt{4356} = 33$;
$9555 \div \sqrt{5402.25} = 130$

Page 72
Slim Sally's CB aerial
Each guy rope is 6.7m long.

Walking puzzle
Lee has walked 13.9km.

Star warrior puzzle
It takes 6.5 minutes for the chamber to fill with water, so the rescuers arrive just in time.

Page 74
Walking the equator puzzle
It would take just over 80 million steps to walk the equator.

Bicycle puzzle
The bicycle wheel must turn 482 times to travel 1km. If it turns 120 times each minute its speed is 15kph.

To work out these two answers you need to find the distance the wheel travels in 1 turn, that is, its circumference. Divide by 100 000 to convert the distance to km, then divide the result into 1 to find the number of turns necessary to travel 1km.

To work out the bike's speed if the wheel turns 120 times each minute, multiply 120 by 60 to find out the number of turns in an hour, then divide by 482 (the number of turns in 1km) to work out the speed in km/h.

Petrol thieves puzzle
The thieves cannot get away in time. It takes almost 11 minutes to fill the 50 cans with petrol.

To work this out, you need to find the volume of one can (in cubic metres) and multiply by 1000 to convert it to litres. Then multiply by 50 to find the total amount of petrol they need, and divide the answer by the rate of flow (900 litres per minute) to find out how long it takes to fill the cans.

Page 75
Bubble puzzle
The surface area of the bubble is 1 263mm².

One way to work it out is to find the radius of the bubble by dividing the circumference by 2π, then square the answer and multiply by 4π to find the surface area.

Page 76
The calculation with missing numbers should be $93 \times 86 = 7998$

Page 78
Trying out the brackets keys
1. 7.3; **2.** 20; **3.** 58; **4.** 94; **5.** 5; **6.** 49; **7.** 770; **8.** 240; **9.** 153

Page 79
More calculations with brackets
1. 11; **2.** 90 006; **3.** 3.03

Page 81
Yachtsman puzzle
The yacht is 89.4m from the lighthouse. The height of the cliff and lighthouse is the side opposite the angle of 40° and the yacht's distance from the headland is the side adjacent to the angle. So you can work out the distance using the formula

$$\tan 40° = \frac{\text{opposite side}}{\text{adjacent side}}$$

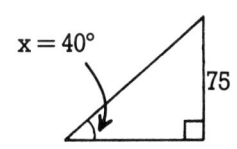

$x = 40°$, 75

$\tan 40° = \dfrac{75}{\text{distance}}$, so distance $= \dfrac{75}{\tan 40°}$

Page 82
Using the power key
$9^8 = 43\,046\,721$; $15^3 = 3\,375$; $6^1 = 6$; $12^4 = 20\,736$; $5^7 = 78\,125$

Negative powers
$3^{-4} = 0.0123457$; $5^{-2} = 0.04$; $4^{-1} = 0.25$; $6^{-5} = 0.0001286$

Working out roots
$\sqrt[8]{6\,561} = 3$; $\sqrt[7]{16\,384} = 4$; $\sqrt[5]{3\,125} = 5$

Page 83
Taking risks
The chances of getting five heads in a row are 0.03125, that is approximately $3/100$, or three in every hundred attempts.

Pages 84-85
Sums in scientific notation
1. 6.8869×10^{17}
2. 1.72×10^{16}
3. 3.3672×10^{17}

Pages 84-85
Sun puzzle
It takes 500 seconds (that is just over 8 minutes) for light from the Sun to reach Earth.

Snail puzzle
It would take the snail 87 671 years to reach the Moon. That is 7.68×10^8 hours divided by the number of hours in a year (24×365).

Paper folding puzzle
The paper would be 1.6888×10^8km high. That is more than 168 million km – right out into space, beyond the Sun.

You can work out 0.15×2^{50} using the power key, or if you want to see how the height of the paper increases, you can enter $\times 2$ in the constant, then enter 0.15 and press = fifty times. Both methods give you the answer in mm, so divide by 1 000 000 to convert it to km.

Page 89
Picking teams
You could choose 1 365 different boys' teams and 167 960 different girls' teams.

Pages 90-92
Statistics puzzle
1. The average number of people at a bus-stop is 6, with a standard deviation of 4.6 people.

2. The average length of time waited by each queue is 17.7 minutes with a standard deviation of 13.7 minutes.

3. Of the people in the picture 58% ($18/31$) are travelling by bus and 13% ($4/31$) are travelling by car.

4. The number of different combinations of people who could sit in the seat is 330, that is $11 \times 10 \times 9 \times 8 \div 4!$

5. There are 6 possible ways of arranging the three figures in the statue.

Calculator power

All calculators run on electrical energy, but there are several different ways of obtaining this energy.

Most calculators use batteries but an increasing number are solar-powered. Also many larger calculators obtain their energy directly from the mains power supply. On this page you can find out a bit more about each of these power sources, and their advantages and disadvantages.

1. Batteries

There are three different kinds of battery which are used to power calculators: the pen light AA battery, the silver oxide battery (otherwise known as the button cell) and the lithium battery.

a) Pen light AA batteries are mainly used in older styles of calculators. They are slightly cheaper than the other two. However, the newer designs in calculators tend not to use them since they are too bulky for very small calculators, and you need at least two of them as a power source.

b) A silver oxide battery, otherwise known as a button cell is, on the other hand, tiny. Its diameter is about the size of a shirt button (hence its nickname), and it is about 0.5cm thick. It would last an average calculator user well over a year, and you would need two or three of them to run your calculator. Although more expensive than the pen light AA, it is slightly cheaper than the lithium battery.

c) Although a lithium battery has a larger diameter than a button cell (it is about 2.5cm wide), it is only 2mm thick. Because of this, manufacturers have been able to produce calculators which are as thin as a credit card. It is the most expensive of the three kinds of battery, but you normally need only one to power a calculator and it has the same life expectancy as a button cell.

2. Solar cells

The solar-powered calculator obtains its energy from any form of light (e.g. the sun, a light bulb, or a torch). A solar cell converts light energy into electrical energy which is then used to power the calculator.

The efficiency of the solar cell is measured in lux power, which indicates its sensitivity to light. When solar-powered calculators were first introduced, most solar cells were 150 lux, but now they are only 50 lux. This means that they need only a very small amount of light to obtain enough energy to run a calculator.

Manufacturers are now working on 30 lux solar cells which would run a calculator on just candle light, but it is unlikely that they will try to produce yet more powerful solar cells for calculators as in less light the user would find it very difficult to press the correct keys!

3. Mains power supply

The third power source is the mains, which is used mainly for desk top calculators. Pocket calculators sometimes have optional mains adaptors, but many people find these too restrictive as they prevent them from carrying the calculator around.

CALCULATOR PUZZLES

Nigel Langdon

Edited by Helen Davies
Designed by Graham Round

Illustrated by Naomi Reed, Martin Newton, Rob McCaig,
Mark Longworth and Graham Round

Contents

- 100 Keyboard puzzles
- 102 Finding out about numbers
- 104 Spotting mistakes
- 106 Divisions
- 108 Fraction puzzles
- 110 Being too accurate
- 112 Use your memory
- 114 Upside-down planet puzzle
- 115 Constant calculations
- 116 Percentage puzzles
- 118 Square numbers
- 120 Opposites
- 122 Circular calculations
- 124 Large number problems
- 126 Sums inside sums
- 128 Power puzzles
- 130 More about large numbers
- 132 Statistics puzzles
- 134 Likely or unlikely?
- 136 Puzzles with triangles
- 138 Puzzle answers
- 144 Index

This section contains lots of puzzles to help you practise your calculator skills. It starts with simple puzzles and tricks which will improve your speed and accuracy. Then you can move on to more advanced problems, involving powers, roots, statistics and trigonometry, and there are puzzles to do on a scientific calculator. Throughout this section you will find simple explanations of the maths you need to do the calculations. All the answers are given at the back on pages 138 to 143.

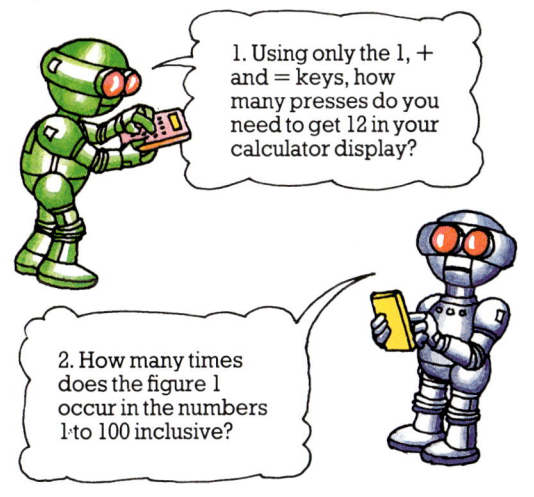

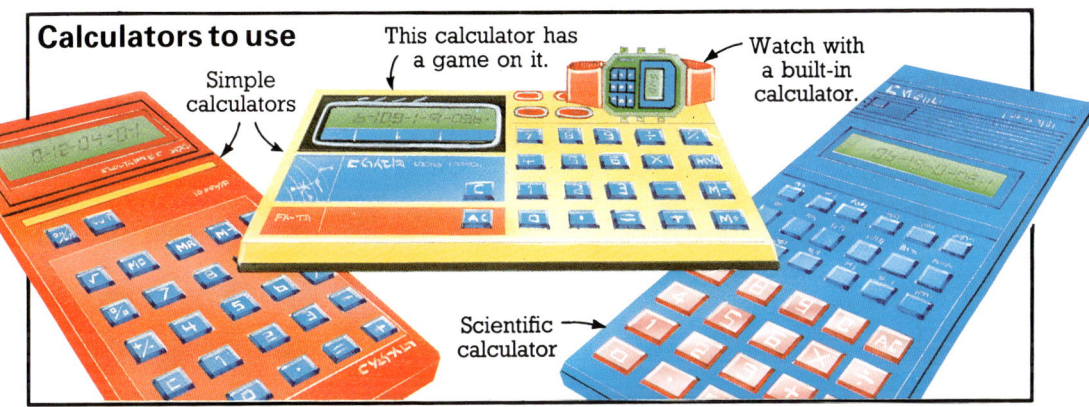

1. Using only the 1, + and = keys, how many presses do you need to get 12 in your calculator display?

2. How many times does the figure 1 occur in the numbers 1 to 100 inclusive?

Calculators to use

Simple calculators

This calculator has a game on it.

Watch with a built-in calculator.

Scientific calculator

You can tackle most of the puzzles in this section using a simple or scientific calculator. Scientific calculators have many more keys than simple calculators, but most of these just provide short cuts, doing jobs which you could do yourself on a simple calculator, such as calculating powers or entering sums with brackets.*

The operations a calculator can do (adding, subtracting and so on) are sometimes called functions. The symbols for the functions vary from one calculator to another. If the symbols on yours are different from those in this book, check them in the instruction manual that came with the calculator.

On scientific calculators, opposite or "inverse" functions, such as squaring and finding square roots, are put on the same key. To select the second function you press a key marked INV (short for inverse). The labels for functions are often colour coded so you can tell which is selected by the INV key.

*Go to pages 50-51 for more about the sorts of calculators you can buy.

Keyboard puzzles

The puzzles on these two pages will help you get familiar with the positions of the basic keys on a calculator.

When you enter a calculation make sure you press the keys firmly and quickly. If your finger hovers too long over a key you may enter a number twice. If you do enter a wrong number you can correct it by pressing the Clear Entry key. This wipes out only the number you have just entered so you do not have to start the whole sum again. Before starting a new calculation always press the All Clear key to erase the whole of the last sum from the calculator.

2. See if you can make the calculator display 100 by pressing only these keys.

3   7   +   −   =

Can you do it with only ten presses?

3. Try to make 1 001 appear in the display using only these keys.

2   7   ×   −   =

How many presses did it take you? A good score is less than ten.

4. Six whole numbers will divide exactly into 1 001. Can you work out what they are?

5. Can you work out which operation keys (+, −, × or ÷) were pressed in this sum?

87 ? 19 ? 31 = 2 108

6. Which numbers are missing in this calculation?

48 × 7? = ? 504

1. What is the sum of all the numbers from 1 to 20 (1 + 2 + 3 and so on)? If you have a watch, time yourself to see how quickly you can do it.

Subtraction puzzles

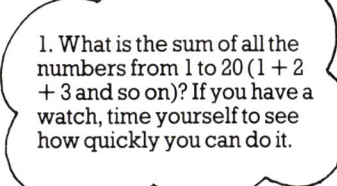

963 − 852 =
852 − 741 =

Try reading the digits on the keyboard downwards and subtracting one column from the next as shown above. What answers do you get?

789 − 456 =
456 − 123 =
Try it backwards too e.g. 987 − 654.

Now read the digits across and subtract one row from the next. Why do you think the answer is the same in all the sums?

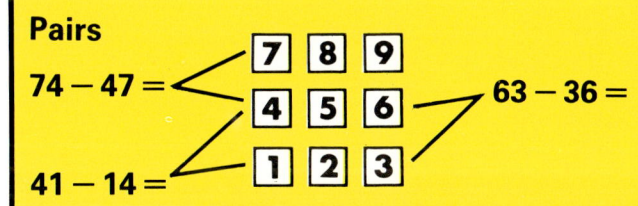

Pairs

74 − 47 =

63 − 36 =

41 − 14 =

Pick any digit on the keyboard and the one below or above it. Reverse the digits to make two numbers, then subtract the smaller from the larger as shown here. Try it several times with different digits and see what happens.

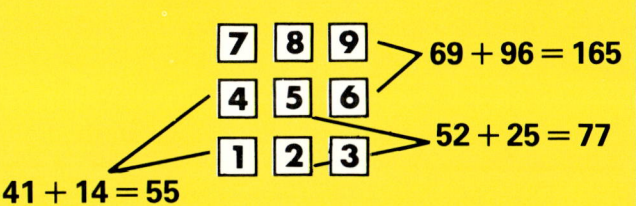

69 + 96 = 165

52 + 25 = 77

41 + 14 = 55

If you add the two numbers the answer is always a multiple of... (which number?)

Neighbours

How many of the numbers from 1 to 20 can you make, using only two neighbouring number keys and one operation key, as shown below? You are not allowed to use keys which are diagonal neighbours.

5 − 4 = 1   6 ÷ 3 = 2   7 − 4 = 3

Which five numbers cannot be made? If you use keys which are diagonal neighbours can you make more of the numbers? How many numbers still cannot be made?

7. Can you work out which numbers are missing in this sum?
?3 × 8? = 7??8
There are two possible answers. Can you find them both?

101

Finding out about numbers

Our number system is called a decimal or "base 10" system because it has ten digits (0 1 2 3 4 5 6 7 8 9). Using these ten digits we can make any number including fractions. (This is what makes pocket calculators possible.) For numbers over 9 we use combinations of two or more digits, and we can make any size of number because we can change the value of a digit by changing its position.* Look at the numbers below.

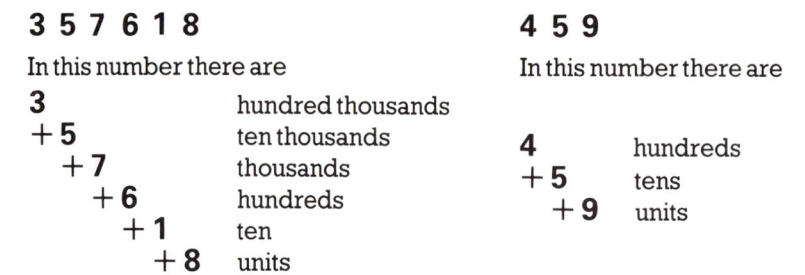

3 5 7 6 1 8

In this number there are

3   hundred thousands
+5   ten thousands
+7   thousands
+6   hundreds
+1   ten
+8   units

4 5 9

In this number there are

4   hundreds
+5   tens
+9   units

In the number on the left the figure 5 represents 50 000 and in the righthand number 5 represents 50. Try the puzzle below on a calculator.

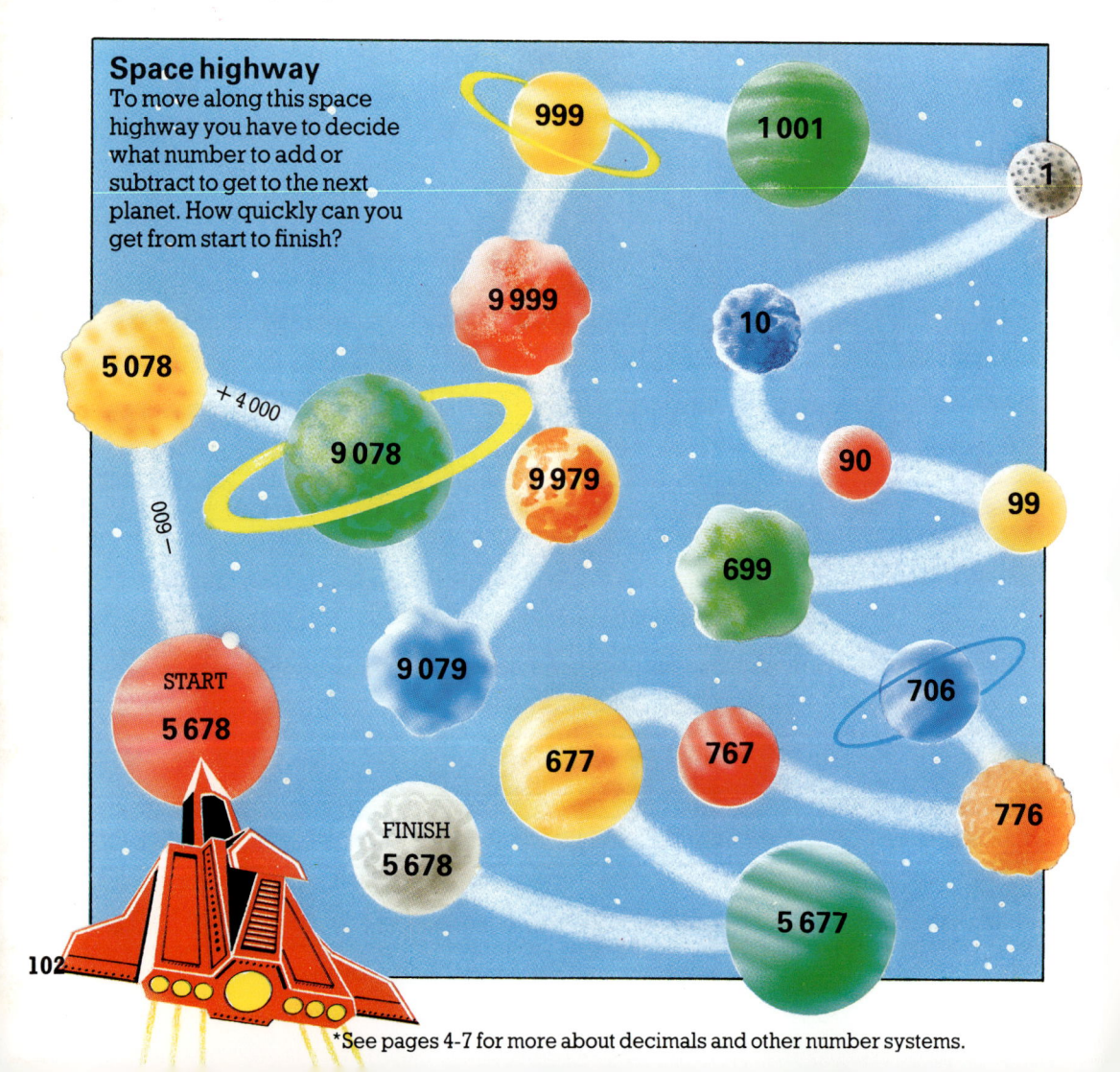

Space highway

To move along this space highway you have to decide what number to add or subtract to get to the next planet. How quickly can you get from start to finish?

999
1 001
1
5 078
+ 4 000
9 999
10
9 078
9 979
90
99
699
9 079
706
START
5 678
677
767
776
FINISH
5 678
5 677

*See pages 4-7 for more about decimals and other number systems.

1 What is it worth?

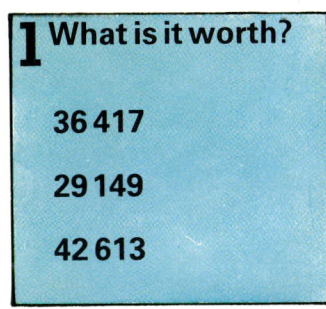

36 417

29 149

42 613

What value does the figure 4 have in each of these numbers?

2

4 762

4 062

What is the difference between these two numbers?

3

472

402

What is the difference between these numbers?

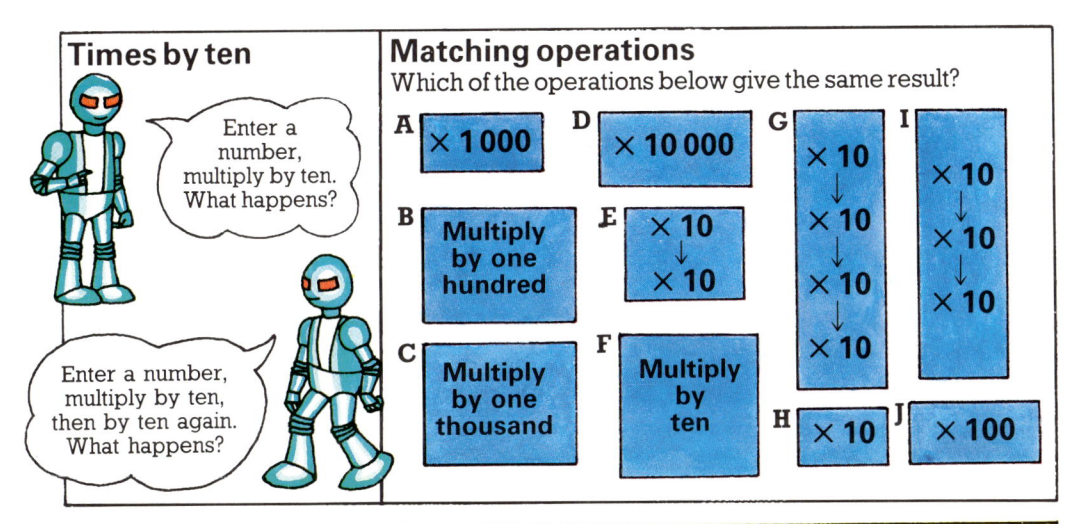

Times by ten

Enter a number, multiply by ten. What happens?

Enter a number, multiply by ten, then by ten again. What happens?

Matching operations
Which of the operations below give the same result?

A × 1 000

B Multiply by one hundred

C Multiply by one thousand

D × 10 000

E × 10 ↓ × 10

F Multiply by ten

G × 10 ↓ × 10 ↓ × 10 ↓ × 10

I × 10 ↓ × 10 ↓ × 10

H × 10

J × 100

Give and take game

Here is a game to play with a friend. The aim is to make your calculator display a number larger than a million (1 000 000).

To play, each person enters a six-figure number into their calculator display (each figure should be different). Then they take it in turns to call out a figure between 1 and 9. The value of the figure in the other player's number is added to the caller's own number, and subtracted from the other player's number, as shown in the example below. First over a million wins.

1st player

216743

216803

216803

2nd player

845167

845107

845107

Give me your 6s

You get 60

Give me your 4s

You get nothing

Give me your 7s

Spotting mistakes

It is very easy to press a wrong key on a calculator without realizing it, so you need to be able to recognize when an answer is not correct. The best way to detect errors is to have a rough idea in your head of what the answer should be. The puzzles on these two pages will give you some practice at knowing what answer to expect.

Which is closest?
Without doing the calculations, what size of answers would you expect to the sums below? Choose A, B or C.

1. **97 × 49**
   A. About 5 000
   B. About 2 000
   C. About 150

2. **36 912 ÷ 12**
   A. About 30
   B. About 300
   C. About 3 000

3. **11 545 − 8 317 + 239 − 1 798**
   A. Between 10 000 and 11 000
   B. Between 3 000 and 4 000
   C. Between 1 000 and 2 000

Find the mistakes
The answer 2 961 is correct for only one of these calculations. Can you work out what mistake was made in each of the others?

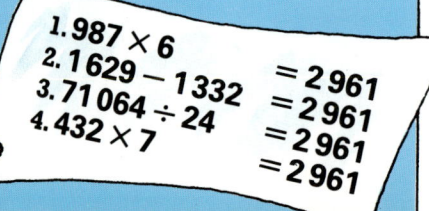

1. 987 × 6 = 2 961
2. 1 629 − 1 332 = 2 961
3. 71 064 ÷ 24 = 2 961
4. 432 × 7 = 2 961

Race to 1
This is a game for one or two players. Choose any starting number (for example, 28) and enter it. Then choose a key number (for example, 3). Using any operation key but only that key number, see how many steps it takes you to reach 1.

Example → 28
[+][3][=] 31
[×][3][=] 93
[+][3][=] 96
[+][3][=] 99
[÷][3][3][=] 3
[÷][3][=] 1

In this example it took six steps to reach 1. Can you do it in less with the same numbers? (It can be done in three steps.)

Starting number → 55   40   27
Key number → [6] [5] [7]

Try the game again with these numbers. If you can get to 1 in eight or nine steps that's good, six or seven is very good and five or less is excellent.

Down to zero
Choose any four-figure number and enter it into your calculator. See if you can reduce the number to zero in exactly four steps. At each step you may only add, subtract, multiply or divide with a two-figure number.

Example → 5327
[−][2][7][=] 5300
[÷][5][3][=] 100
[−][5][0][=] 50
[−][5][0][=] 0

Do you think all four-figure numbers can be reduced to zero in four steps?

Follow-ons

1. The number 66 can be made by adding four consecutive* numbers. What are they?

2. The number 1 190 is the result of multiplying two consecutive numbers. What are they?

3. The number 504 is the result of multiplying three consecutive numbers. Can you work out what they are?

*Consecutive numbers are ones which directly follow one another e.g. 8, 9, 10.

Reversing puzzle

Enter two digits into the calculator display, the largest first.

`62`

Repeat them to make a six-figure number.

`626262`

Then work out what number you need to subtract to give an answer which contains the same digits in reverse.

Need to subtract 363636

`262626`

Try the puzzle with lots of different pairs of digits. What do you notice about the numbers you have to subtract?

Four-in-a-line game

This game is for two or more players. Take it in turns to choose two numbers from the green panel and multiply them. Then cover the answer on the board below with a coin or a piece of paper. The winner is the first person to cover four numbers in a line.

8	31	12
20	5	71
51	63	22

Board numbers:
1122, 176, 440, 355, 3621, 568, 408, 852, 315, 620, 1020, 612, 2201, 110, 1581, 248, 60, 40, 1420, 4473, 682, 255, 756, 155, 96, 1386, 1953, 1260, 504, 160, 264, 3213, 372, 240, 1562

105

Divisions and decimals

Division trick

Write down a three-figure number. Then enter it twice into a calculator to make a six-figure number in the calculator display.

358358

Divide by 11, then by 13, then by 7. What happens?

Try the trick with lots of three-figure numbers.

How it works

Repeating the figures of a three-figure number is the same as multiplying it by 1 001.

358 × 1 001 = 358 358

358 × 1 001 is the same as 358 × 1 000 plus 358 × 1.

Dividing by 11, then 13, then 7 is the same as dividing by 1 001 because 11 × 13 × 7 = 1 001.

358 358 ÷ 1 001 = 358

The numbers 11, 13 and 7 are called factors of 1 001 because they are whole numbers which, when multiplied together, make 1 001.

Can you think up a four-digit trick which uses the fact that 73 and 137 are factors of 10 001?

Leftovers

If you divide a number by another number which is not one of its factors, the answer will not be a whole number. It can be expressed in several different ways. Look at the example below.

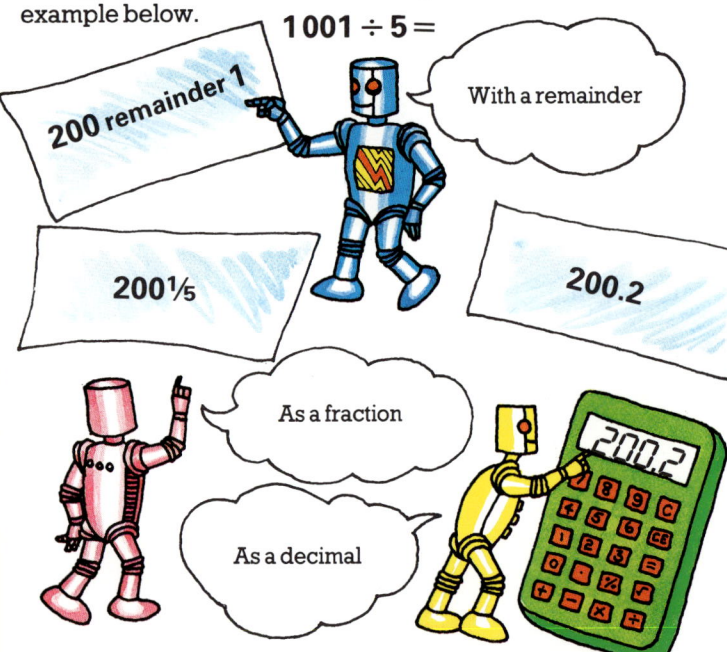

1 001 ÷ 5 =

200 remainder 1 — With a remainder

200 1/5 — As a fraction

200.2 — As a decimal

A calculator gives the answer as a decimal. It cannot show remainders or fractions. If a number does not divide exactly, the calculator carries on dividing into the remainder. Can you think of a way to work out the remainder from the calculator's answer? The puzzle below may give you a hint.

Matching puzzle

Can you match these divisions with the answers in the displays? Use a calculator to check your results.

17 ÷ 2
107 ÷ 10
100 ÷ 8
37 ÷ 4
8 ÷ 16
12 ÷ 48
54 ÷ 12
10 ÷ 8

0.5
4.5
10.7
12.5
0.25
8.5
9.25
1.25

Endless answers

20 ÷ 12 = 1.666 666 666 666 666

Often, even the remainder will not divide exactly. This produces a "recurring" decimal in which the numbers after the point are repeated endlessly.

`1.6666667`

Most calculators have space for eight digits in the display, so you never see more than seven of the recurring digits. On some calculators the last digit in the display is rounded off.

1 Which is bigger?

0.066 666 6

0.2

Which of these two numbers is bigger? Remember, it is the position of the figures as well as their size, that is important.

More matching puzzles

Here are some more divisions to match with their answers.

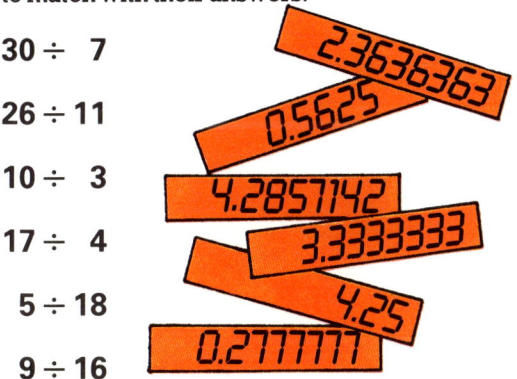

30 ÷ 7
26 ÷ 11
10 ÷ 3
17 ÷ 4
5 ÷ 18
9 ÷ 16

2.3636363
0.5625
4.2857142
3.3333333
4.25
0.2777777

2

0.066 666 6 is 1/15 or, 1 ÷ 15

0.2 is 1/5 or, 1 ÷ 5

In the number 0.066 666 6 the first figure after the decimal point is a zero. In the number 0.2 the first figure after the point is a 2, so 0.2 is the bigger number.

3

AN
AND
ANORAK
ANT
ANTARCTIC
APE
APOCALYPSE

APPROPRIATE
ART
ARTIST
AS

1.89
1.9073
1.9352

2.1
2.15
2.166 666
2.25
2.4
2.43

A good way to sort decimal numbers into order of size is to think of them like words in a dictionary. For example, AN comes before APOCALYPSE because N comes before P in the alphabet.

4

2.75
C
2.5
2.474 747 4
2.4
B
2.25
A
2.166 666 6
2.15
2.1

These cards are in numerical order. The three blank cards hold the answers to the calculations 7 ÷ 3, 49 ÷ 19 and 440 ÷ 200. Can you work out which calculation belongs to which card?

Fraction puzzles

To work out fractions on a calculator you have to convert them to decimals by dividing the top part by the bottom part as shown below.

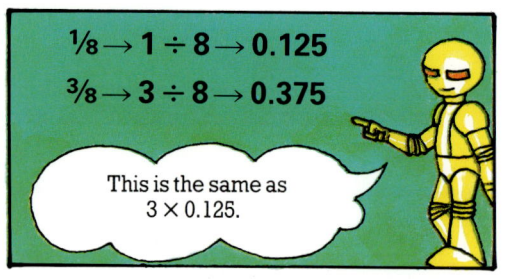

$1/8 \rightarrow 1 \div 8 \rightarrow 0.125$
$3/8 \rightarrow 3 \div 8 \rightarrow 0.375$

This is the same as 3×0.125.

Some fractions produce interesting results. Try $1/3$, $2/3$ and $3/3$.

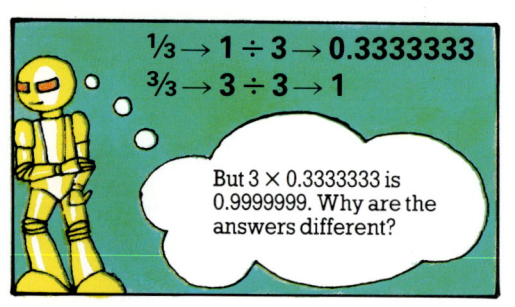

$1/3 \rightarrow 1 \div 3 \rightarrow 0.3333333$
$3/3 \rightarrow 3 \div 3 \rightarrow 1$

But 3×0.3333333 is 0.9999999. Why are the answers different?

The reason for the different answers is that 0.3333333 is not exactly $1/3$, but it is the nearest you can get on a calculator. When it is multiplied by 3 the answer is very close to, but not exactly, 1.

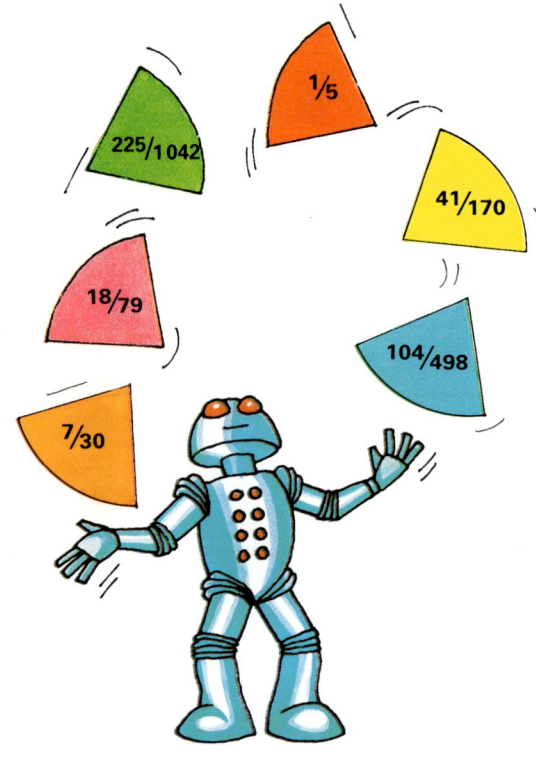

Fractions shown: $225/1042$, $1/5$, $41/170$, $18/79$, $104/498$, $7/30$

Sizing up fractions

It is often very difficult to know whether one fraction is bigger than another, especially when they have different figures at the bottom. If you convert the fractions to decimals it is much easier to tell. Can you arrange the above fractions in order of size, smallest first?

Triad

This is a game to play with a friend. You need a piece of paper and two different coloured pens or pencils.

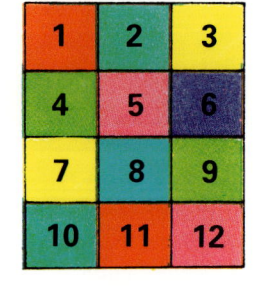

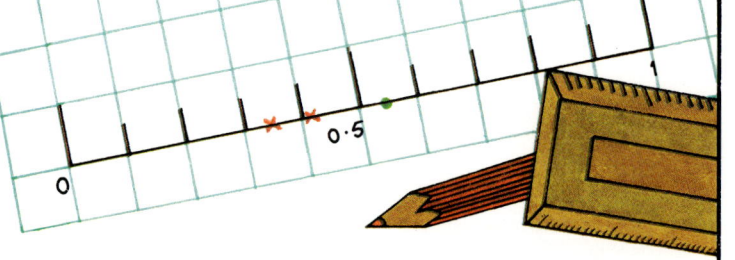

1	2	3
4	5	6
7	8	9
10	11	12

Draw a line on the paper and divide it into tenths. Label the ends of the line 0 and 1 and the centre-point 0.5.

To play the game each player in turn chooses two numbers from the chart on the left to make a fraction, then converts it to a decimal and marks its place on the line. The aim of the game is to get three marks on the line without any of your opponent's marks in between.

If the decimal equivalent of your fraction is more than 1, your mark goes off the line and you miss a go.

Lost fraction puzzle

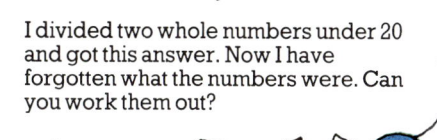

I divided two whole numbers under 20 and got this answer. Now I have forgotten what the numbers were. Can you work them out?

0.2307692

1 Finding patterns

$1/7$ $2/7$ $3/7$
$4/7$ $5/7$ $6/7$

If you convert sevenths fractions to decimals you get the same pattern of digits starting in a different place in each answer. Try converting the sevenths shown above and see if you recognize the pattern.

Tops and bottoms

$5/10 = 0.5$

$6/11 =$

Is $6/11$ more or less than 0.5?

2

$1/7$

First recurring digit
0.1428571

What would be the first twelve digits for $1/7$?

In fact, each of the sevenths fractions produces a recurring decimal in which a pattern of six digits repeats itself. Most calculator displays only have room to show the pattern once but you can see the first recurring digit.

3

$1/17 = 0.0588235$

$2/17 = 0.1176470$

$3/17 = 0.1764705$

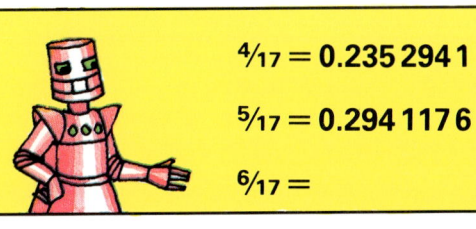

$4/17 = 0.2352941$

$5/17 = 0.2941176$

$6/17 =$

Seventeenths fractions produce a recurring pattern with sixteen digits. This is much too long for a calculator to display. Can you find the pattern from the conversions shown above and work out all sixteen recurring digits for $1/17$?

Try to predict the decimal value of $6/17$ without using a calculator.

Can you find any other interesting patterns when converting fractions to decimals? Elevenths and thirteenths are good ones to try.

Fractions made easy

$3/10 \times 2\frac{1}{2}$ $3/5 \times 4/7$ $7/8 + 3/4$

$1/2 \div 1/4$ $5\frac{1}{8} - 3\frac{3}{4}$ $1/2 \times 1/2$

Calculations with fractions are much easier to do using a calculator. You just change the fractions to decimals before you start. Try these examples.

To save writing down the decimal equivalents of the fractions, you could store one of them in the memory while you work out the other.

109

Being too accurate

With an eight-digit display calculators are often much more accurate than you need. For instance, say you wanted to find out how long it would take to travel a journey of 270km at 110km/h.

270 ÷ 110 = 2.454 545 4 hours

The calculator's answer is not a very useful reply to the question because it is too detailed. To plan the journey all you need to know is that it would take roughly two and a half hours. So rounding off 2.454 545 4 to 2.5 is accurate enough for this particular question.

When you solve a problem you should think about how accurate you need to be to give an answer which is meaningful for the question. How would you round off the numbers in the statements below if you were telling them to a friend?

1. A light year is 5 865 696 000 000 miles or 9 385 113 600 000 kilometres.

2. There are 30 126 541 cats in the USA.

3. The population of London is 6 877 142.

4. The shortest street in Britain is 17.672 metres long.

Puzzles

1. The world speed record for the fastest train is 410 kilometres per hour. In 1829 the record was 29.1 miles per hour. How many times greater is the new record? (1 mile is $8/5$ kilometres.)

2. The longest railway line in the world stretches 9 438km from Moscow to Nakhodka. How long would it take a train travelling at an average speed of 120km/h to complete the journey?

3. Have you lived a million hours?

Three-figure accuracy

To round off the distance 293 467km to the nearest ten thousand kilometres gives 290 000km. Only the first two figures in the number are accurate and these are called the significant figures.

As a general rule an answer with *three* significant figures is accurate enough for most problems. Rounding a number to three significant figures means you are ignoring a very small part of the answer. You can see this in the example below.

293 467 becomes **293 000**

The error is **467** in **293 000** or $467/293\,000$

That is about $500/300\,000$ or $1/600$.

Rounding the distance 293 467km to three significant figures gives an error of 467 in 293 000 or about 1 in 600, so you are only ignoring one six-hundredth of the answer.

Hints for rounding off

1. The zeros between the significant figures and the decimal point do not count.

0.005 213 3 → 0.005 21

26 833 000 → 26 800 000

2. If the fourth figure is a 5 or above you should add 1 to the third significant figure.

0.005 216 → 0.005 22

543.724 → 544

Can you round these answers to three significant figures?

1. 0.006 376 3
2. 0.062 871 9
3. 264.374 12
4. 781 432.16
5. 0.199 999

Space puzzles

When you do these puzzles, round off the answers so they are meaningful to you.

1. The Moon is 240 000 miles away. How long would a spaceship take to travel to the Moon at 1 mile per second?

2. A spaceship leaves Earth and travels at 40 000km/h. How long will it take to reach Mars which is 56 million km away?

3. The Sun and the Moon appear about the same size. But the diameter of the Sun is 1 382 400km and the diameter of the Moon is only 3 480km. How many times wider is the Sun than the Moon really?

In-between game

To play this game you need two people. Take turns to choose two numbers from the chart on the right. Divide one by the other, then check the scoreboard to see how many points you have scored. The first player to score 10 points wins. (A number cannot be used twice in one game.)

SCOREBOARD

Answer	Score
Between 0 and 1	1 point
Between 1 and 10	2 points
Between 10 and 100	3 points
Over 100	zero

CHART

9	23	31	46
97	129	152	216
255	364	440	800
1974	2132	2561	2619
2815	3966	4770	9342
13 000	14 500	16 000	29 500

Use your memory

The memory on a calculator is very useful when you are doing calculations with several parts. The labels on the memory keys vary from one calculator to another and if yours are different from the ones shown here check your instruction manual.

To store a number in the memory you use the M+ key. On many calculators you erase a number from the memory by pressing the MR (memory recall) key twice. On others you press MR to display the number, then M− to subtract the number in the display from the one in the memory.

Try these calculations to practise using the memory keys.

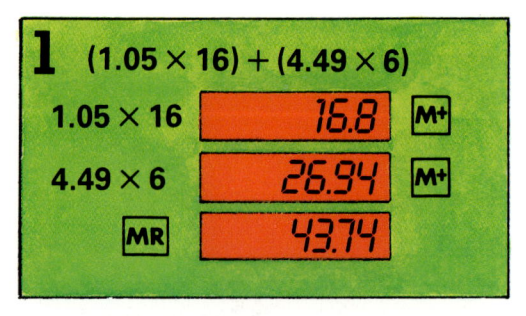

1 (1.05 × 16) + (4.49 × 6)

1.05 × 16	16.8	M+
4.49 × 6	26.94	M+
MR	43.74	

To work out this sum you do the first multiplication and store the answer in the memory. Then do the second multiplication and add the result to the memory by pressing the M+ key. The memory now holds the total of both sums. To display it press the MR key.

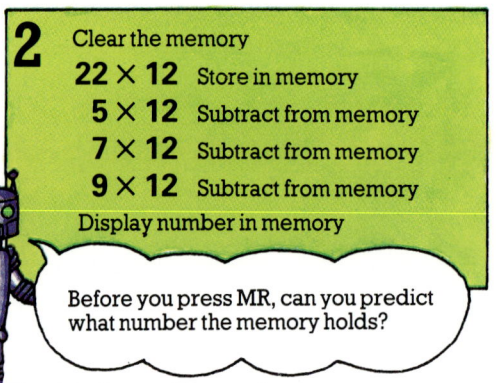

2 Clear the memory
22 × 12   Store in memory
5 × 12   Subtract from memory
7 × 12   Subtract from memory
9 × 12   Subtract from memory
Display number in memory

Before you press MR, can you predict what number the memory holds?

The M− key subtracts the number in the display from the one in the memory. Try the above routine to practise using it. Leave the final answer in the memory so you can use it in the next routine.

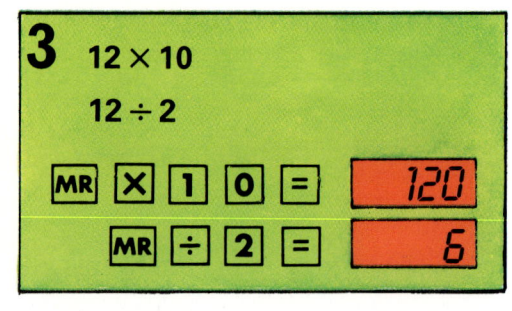

3 12 × 10
12 ÷ 2

| MR | × | 1 | 0 | = | 120 |
| MR | ÷ | 2 | = | 6 |

Once a number is stored in the memory you can use it for lots of separate calculations. Try the examples above. Using the number in new calculations does not wipe it from the memory. It is only removed if you erase it or switch the calculator off.

Repeats

Here are some calculations which produce interesting number patterns. Each one involves using a number over and over again and you could store that number in the memory.

1
1 × 9 + 2 =
12 × 9 + 3 =
123 × 9 + 4 =
1234 × 9 + 5 =

2
143 × 2 × 7 =
143 × 3 × 7 =
143 × 4 × 7 =
143 × 5 × 7 =
143 × 6 × 7 =

What number did you put in the memory?

3
1 × 8 + 1 =
12 × 8 + 2 =
123 × 8 + 3 =
1234 × 8 + 4 =

See if you can work out why the patterns occur. There are some hints with the answers at the back of the book.

Conversions

| 85km | 100km | 50km | 270km |
| 126km | 32km | 10km | 115km |

Can you convert these distances in kilometres to miles? (A kilometre is ⅝ of a mile.) Try using the memory to speed up the calculations.

Order of calculations
When you are using the memory to work out divisions you have to plan the calculation carefully. Look at the example below.

$$\frac{285 + 117}{264 - 68}$$

1. 264 − 68 = M+
2. 285 + 117 =
3. ÷ MR =

Do the sum in this order.

To do this calculation you need to work out the bottom part first and store the result in the memory. Then work out the top part and divide this answer by the number in the memory. For more practice with calculations like this try the planet puzzle on page 114.

Who wins?
Can you solve this puzzle? You need to plan the calculation carefully so you can use the memory on your calculator. The horse can run a mile in 1 minute and 35 seconds, and the woman can run 800 metres in 1 minute 44 seconds. If the woman is given a 1000 yards start, who will win the mile race?

Hints for working it out
You need to work out the woman's running speed in metres per second, and you will also have to convert the distance she still has to run into metres.
(1 mile = 1 760 yards and 1 yard = 0.9144 metres.)

Upside-down planet puzzle

You are the commander of a spaceship which is on a reconnaissance mission of a group of planets. Coded instructions for where to go next are hidden in mathematical calculations which you are given each time you reach a planet. To read the instructions you need to work out the calculations and then decode the answer. Starting from Earth what is the final message you receive?

LESBOS

Next go to

$$5\,000 \times \left[\frac{47\,512.562}{803.5 \times 4} \right]$$

ISIS

SIBEL

Next

$$\frac{95\,380\,000 + 8\,706}{193 \times 7}$$

Next

$$\frac{96.47 + 4\,998.9}{0.007\,64 - 0.001 + 0.003\,36}$$

LILOSI

Fly to

$$\frac{8.3844}{20.63 - 6.93}$$

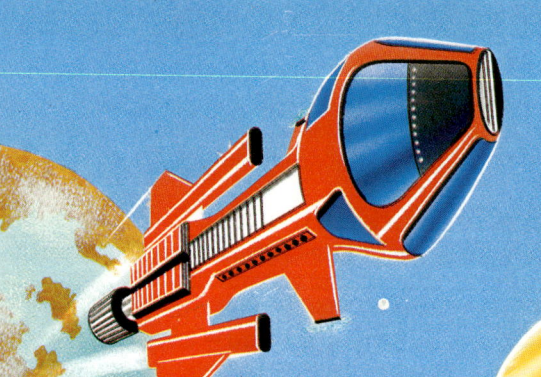

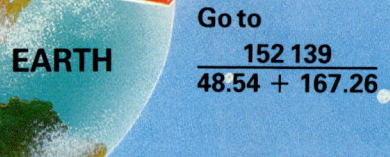

EARTH

Go to

$$\frac{152\,139}{48.54 + 167.26}$$

SOL

Next visit

$$\frac{8\,515\,510.5}{2\,647.64 - 2\,591.14}$$

Now

$$\frac{141.064\,27}{157.8 \div 789}$$

GOGOL

ZIGO

Full speed now to

$$\frac{1717 \times 441}{37\,191 \div 253}$$

114

Constant calculations

The constant function on a calculator is a kind of automatic memory. It allows you to repeat a function and number without re-entering them. Not all calculators have a constant function. To check whether yours has, try this test.

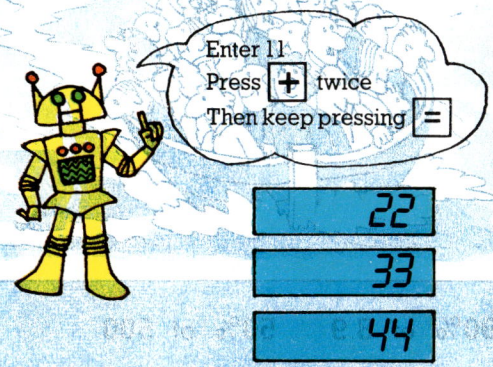

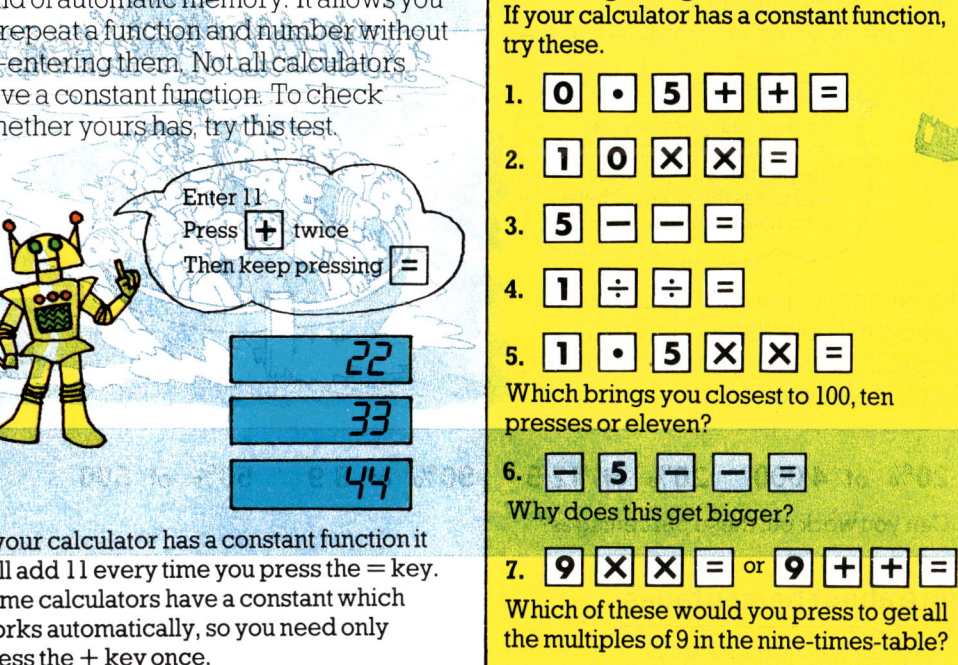

Enter 11
Press [+] twice
Then keep pressing [=]

22
33
44

If your calculator has a constant function it will add 11 every time you press the = key. Some calculators have a constant which works automatically, so you need only press the + key once.

Investigating constants
If your calculator has a constant function, try these.

1. [0] [.] [5] [+] [+] [=]
2. [1] [0] [×] [×] [=]
3. [5] [−] [−] [=]
4. [1] [÷] [÷] [=]
5. [1] [.] [5] [×] [×] [=]

Which brings you closest to 100, ten presses or eleven?

6. [−] [5] [−] [−] [=]

Why does this get bigger?

7. [9] [×] [×] [=] or [9] [+] [+] [=]

Which of these would you press to get all the multiples of 9 in the nine-times-table?

1 Using the constant

60 ÷ 5
5 000 ÷ 5
12.5 ÷ 5
0.5 ÷ 5

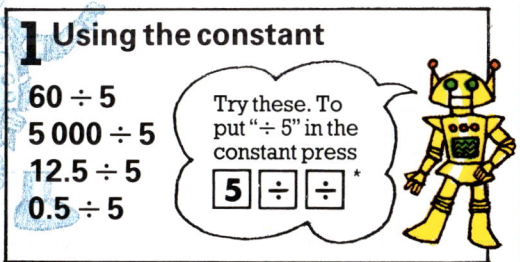

Try these. To put "÷ 5" in the constant press [5] [÷] [÷] *

Once you have given the calculator a constant such as "÷ 5", you can use it on any number you enter, so long as you do not clear the display or press an operation key.

2

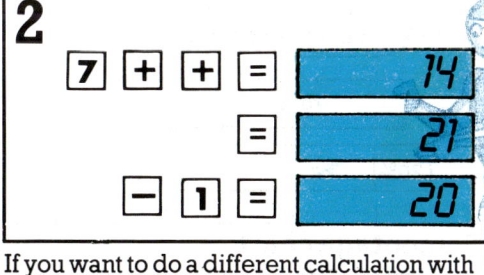

[7] [+] [+] [=]   14
           [=]   21
      [−] [1] [=]   20

If you want to do a different calculation with a number which appears in the display, you can do so because pressing any operation key will cancel the constant.

3

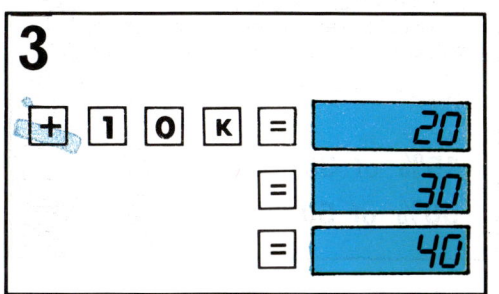

[+] [1] [0] [K] [=]   20
              [=]   30
              [=]   40

Some calculators, particularly scientific ones, have a constant key labelled K. To use the constant you enter the function and number you want to repeat, then press K, as shown above.

If a plant is 1cm high and it doubles in size every day, how tall will it be after 15 days?

*On calculators with an automatic constant you may need to press the 5, ÷ and = keys before starting.

Percentage puzzles

The island of Brigg has a population of 249 000. If 20% of the population emigrate, how many people leave the island?

To work out percentages you multiply the number by the percentage you want and press the % key. There is no need to press =.

The number of people left is now 80% of the original population. How many people is that? Can you think of a way to check the answer?

> 20% of 4 000   20% of 12.5   90% of 8.9   50% of 500
>
> Can you work out these percentages?

More about percentages

A percentage is a useful way to describe a fraction of something. For instance, 20% is the same as 1/5 and 75% is the same as 3/4.

75%

20%

Like fractions, percentages can be expressed as decimals.

20% → 20/100 → 1/5 → 0.2

75% → 75/100 → 3/4 → 0.75

Percentage chart

%	FRACTION (OVER 100)	SIMPLE FRACTION	DECIMAL
50%	50/100		0.5
25%	25/100		
10%		1/10	
33⅓%	33.33/100	1/3	0.3333
15%		3/20	
		2/3	

The chart above shows the most commonly used percentages with their fraction and decimal equivalents. Can you fill in the missing numbers?

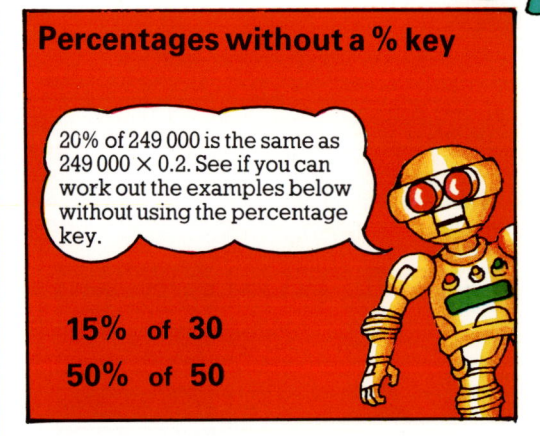

Percentages without a % key

20% of 249 000 is the same as 249 000 × 0.2. See if you can work out the examples below without using the percentage key.

15% of 30
50% of 50

If your calculator does not have a percentage key, you will need to convert percentages to decimals to work them out. To do this you divide by 100 as shown in the chart on the left.

What's the difference?

If the weight of an astronaut from Earth increases by 20% on planet Zardoz, how much would you weigh on Zardoz? You can work this out in six different ways, as shown below. Methods 1, 2 and 3 give the percentage increase, which you then add on to your weight. Methods 4, 5 and 6 give the new weight direct.

1. Multiply weight by 20%.
2. Multiply weight by 20 and divide by 100.
3. Multiply weight by 0.2.
4. Multiply weight by 120%.
5. Multiply weight by 120 and divide by 100.
6. Multiply weight by 1.2.

Half-life

The radioactive output of an imaginary chemical, Zilium, is 463 units, but it decreases by 50% every day. How many days will it take for the radioactive output to be within the "safe" level of 4 units?

Orang-utan puzzle

The orang-utan is in danger of extinction. If there are 5 000 left in the wild now and their numbers decreased by 15% each year, how many years would it be before there are less than 2 500?

Prize money puzzle

You have won a TV quiz, but there is one final question: how do you want to receive the money? There are two choices.

1. You can have 100 banknotes the first year, 10% less the next year, 10% less the year after and so on for ten years.

2. You can have 10 banknotes the first year, 50% more the year after and so on for ten years.

Which method would you choose? (You could use the memory on your calculator to keep a running total of the amounts you get each year.)

Square numbers

Squaring a number means multiplying it by itself. For example, seven squared (written 7^2) is 7×7. The square of 7 is therefore 49.

Small square numbers are easy to calculate in your head but you need to use a calculator to find larger squares.

[2] [5] [×] [=] **625**

0.5^2
37^2

Try these too.

[2] [.] [5] [x^2] **6.25**

On most calculators you can use the constant function to square a number.*

Some calculators have a squaring key marked x^2. This will give you the square of any number you enter.

1 Square number puzzles

12^2
21^2
113^2
311^2
1003^2
3001^2
201^2
102^2

These pairs of square numbers produce interesting results when you work them out. See if you can find any others like them.

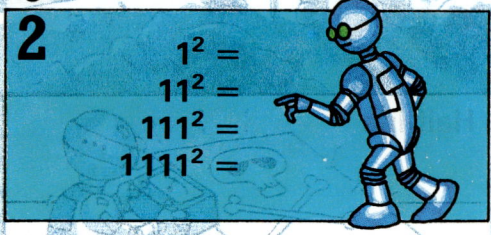

2

$1^2 =$
$11^2 =$
$111^2 =$
$1111^2 =$

Here are some more squares to investigate. What pattern of answers do they produce?

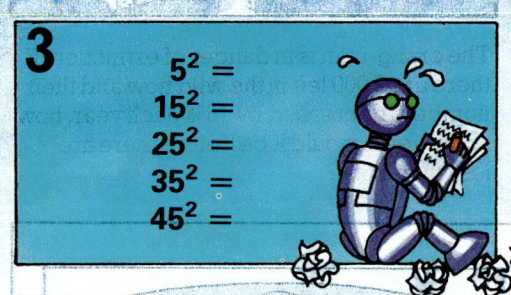

3

$5^2 =$
$15^2 =$
$25^2 =$
$35^2 =$
$45^2 =$

4

$1301 = ?^2 + ?^2$

Work out these squares and write down the answers. Then see if you can work out the square of 55 without using a calculator (or pencil and paper).

The figure 1 301 is the sum of the squares of two consecutive numbers. Can you calculate what they are?

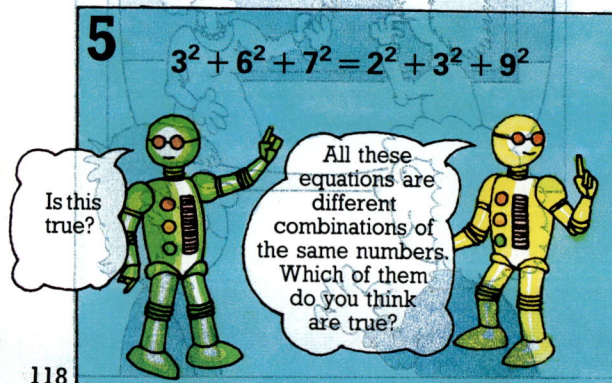

5

$3^2 + 6^2 + 7^2 = 2^2 + 3^2 + 9^2$

Is this true?

All these equations are different combinations of the same numbers. Which of them do you think are true?

1. $32^2 + 63^2 + 79^2 = 23^2 + 36^2 + 97^2$
2. $33^2 + 69^2 + 72^2 = 33^2 + 96^2 + 27^2$
3. $32^2 + 69^2 + 73^2 = 23^2 + 96^2 + 37^2$
4. $39^2 + 62^2 + 73^2 = 93^2 + 26^2 + 37^2$
5. $39^2 + 63^2 + 72^2 = 93^2 + 36^2 + 27^2$
6. $33^2 + 62^2 + 79^2 = 33^2 + 26^2 + 97^2$

*You will probably only need to press the × key once.

Ancient triangles

Measurements at prehistoric sites, such as Stonehenge in England, show that Stone Age builders had discovered the principles of Pythagoras' theorem* long before Pythagoras was born.

A theorem is a mathematical statement which can be proved to be true

Pythagoras' theorem states that whenever a triangle has a square corner (that is, a right-angle of 90°), the square of the longest side is equal to the sum of the squares of the other two sides. You can check this on the triangle shown above.

The theorem also works the other way round. If the squares of two sides of a triangle equal the square of the third side, the triangle must contain a right-angle. Is the above triangle right-angled?

3, 4, 5   12, 35, 37   5, 12, 13   19, 59, 62

41, 71, 82   8, 15, 17   8, 9, 12

The Stone Age builders used triangles in order to construct right-angled corners. However they were not always absolutely accurate. The measurements above are from triangles found in Stonehenge and other sites. Which ones give perfect right-angles?

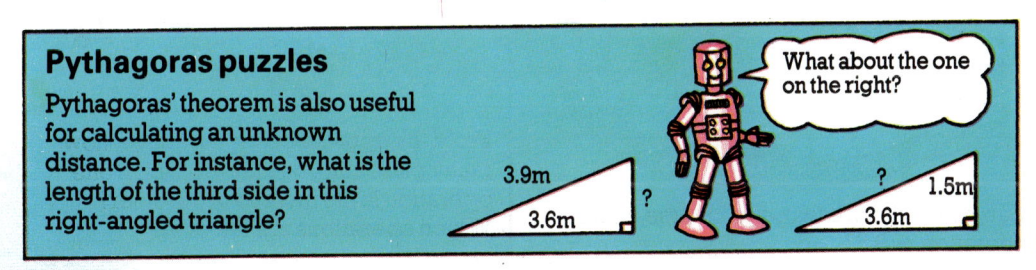

Pythagoras puzzles

Pythagoras' theorem is also useful for calculating an unknown distance. For instance, what is the length of the third side in this right-angled triangle?

3.9m, 3.6m, ?

What about the one on the right?

?, 1.5m, 3.6m

*See pages 27 and 72 for more about Pythagoras' theorem.

Opposites

Some operations on a calculator have the opposite, or inverse, effect of one another. For example, dividing by 3 reverses the effect of multiplying by 3. Look at the example on the right.

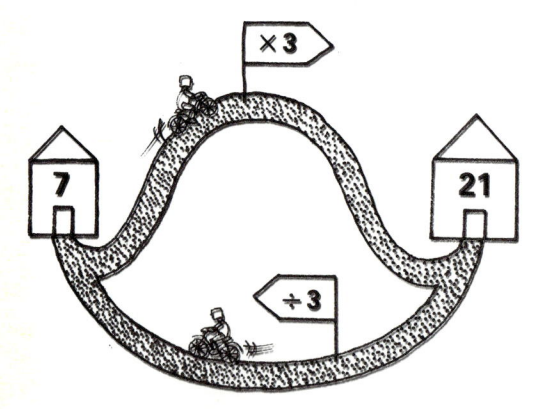

Now try these.

1

The four basic operations on a calculator can be divided into two pairs of inverses. Which are the pairs?

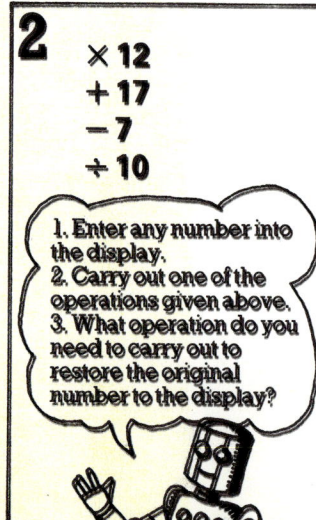

2

× 12
+ 17
− 7
÷ 10

1. Enter any number into the display.
2. Carry out one of the operations given above.
3. What operation do you need to carry out to restore the original number to the display?

Try the robot's routine above to find the inverses of each of the operations at the top of the box.

3

The operation ÷ 3 reverses the effect of × 3, but does × 3 reverse ÷ 3? Try it on several different numbers and see.

Alternatives

There is usually more than one inverse for a particular operation. For example if you multiply by 2, you can get back to the original number by ÷ 2 or × 0.5.

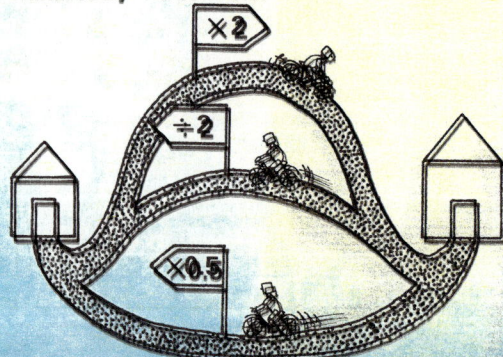

Can you find more than one inverse of the operation ÷ 0.5?

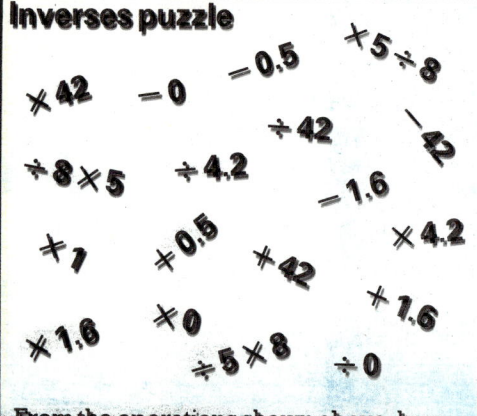

Inverses puzzle

× 42   − 0   − 0.5   × 5   ÷ 8
         ÷ 42   − 42
÷ 8   × 5   ÷ 4.2   − 1.6
× 1   + 0.5   × 4.2
              + 42   + 1.6
× 1.6   × 0   ÷ 0
         ÷ 5   × 8

From the operations shown above, how many pairs can you make in which one operation is the inverse of the other? Which operations are left over?

Square roots

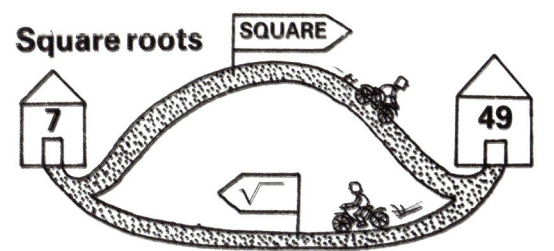

Finding a square root is the inverse of squaring. The square root of 49 (written $\sqrt{49}$) is 7.

Some calculators have a square root key labelled $\sqrt{}$. You can see how to use it above.

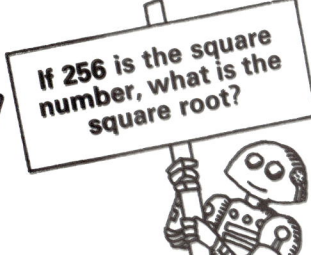

7 $\sqrt{}$ → 2.6457513

If 16 is the square root, what is the square number?

If 16 is the square number, what is the square root?

If 256 is the square number, what is the square root?

Square roots without a $\sqrt{}$ key

Finding a square root without a square root key is a matter of trial and error. For example to find $\sqrt{70}$ first make a guess.

$8^2 = 64$ and $9^2 = 81$, so try 8.5^2.

72.25

Too big, try 8.4^2.

70.56

Still too big, try 8.3^2.

68.89

What would you try next?

Can you continue the calculation to find the square root of 70 correct to three decimal places (i.e. your answer should show three figures after the decimal point)?

In fact a calculator also works out square roots by trial and error, and some calculators take longer to display square roots than other answers.

Square root trick

To do this mind-reading trick ask a friend to think of two numbers with a difference of 2 between them (e.g. 16 and 18) and to remember them without telling you what they are. Then give your friend a calculator and the following instructions:

Multiply the two numbers. — 1 6 $\times$ 1 8 $=$

Then add 1. — $+$ 1 $=$

Now take the calculator and find the square root of the number in the display. — 289

$\sqrt{}$

17

Adding 1 to this result will give you one of the numbers your friend thought of and subtracting 1 from it will give the other. Try the trick with lots of different numbers.

121

Circular calculations

People have known for a long time that, no matter what the size of a circle, its circumference is always "three and a bit" times longer than its diameter. However the exact size of the "bit" has not been found. The number "three and a bit" is represented by the Greek letter π (pronounced pie). Modern computers have calculated the value of π to several thousand decimal places. The first six hundred are shown below.

π = 3.14159265358979323846264338329750288419716939
9375105820974944592307816406286208998628034825342
1170679821480865132823066470938446095505822317253
5940812848111745028410270193852110555964462294895
4930381964428810975665933446128475648233786783165
2712019091456485669234603486104543266482133936072
6024914127372458700660631558817488152092096282295
4091715364367892590360011330530548820466521384146
9519415116094330572703657595919530921861173819326
1179310511854807446237996274956735188575272489122
7938183011949129833673362440656643086021394946395
2247371907021798609437027705392171762931767523846
74818467669405132

Over the centuries, different mathematicians have suggested various values for π. Some of these are shown below. Which one is closest to the computer's figure?

$4 \times (1 - \frac{1}{9})^2$ ← Egyptian (1650 BC)

$3\frac{1}{8}$ ← Babylonian (before 500 BC)

$\sqrt{2} + \sqrt{3}$ ← Greek (450 BC)

$355/113$ ← Chinese (AD 500)

$3927/1250$ ← Indian (AD 400)

Between **$3\frac{1}{7}$** and **$3\frac{10}{71}$** ← Archimedes (220 BC)

Using π

Knowing the value of π enables you to calculate the curved distance around a circle or sphere (that is, its circumference). For instance, this can has a diameter of 12cm, and its circumference is $\pi \times 12$.

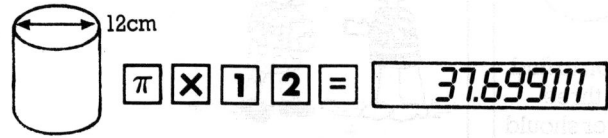

| π | $\times$ | 1 | 2 | = | 37.699111 |

If your calculator has a key which displays the number π you can work out the answer as shown above. (If not, enter a rounded-off version of π, e.g. 3.142.) The can would need a label 37.7cm long to go right round it. How long would the label for a can with a 16cm diameter need to be?

Earth puzzles

1. The diameter of the Earth is 12 640km, how long is its circumference?

2. If a satellite orbits Earth at an altitude of 100km, what is the length of its orbit?

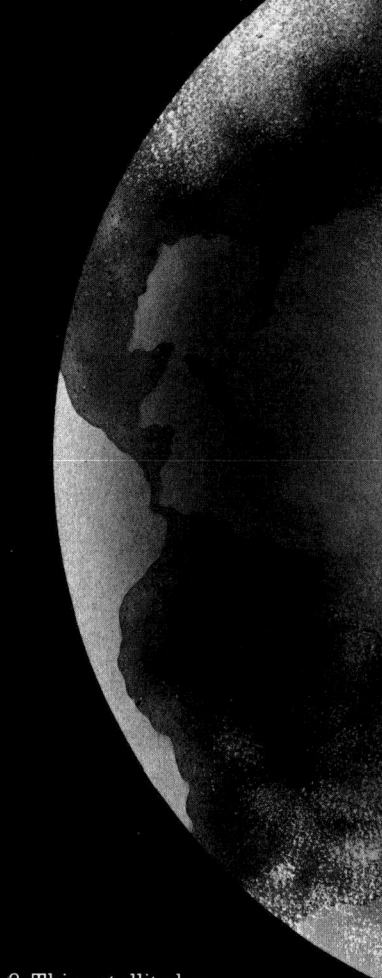

3. This satellite has a diameter of 5m. A signal-reflecting band encircles the outer casing. How long is the band?

4. If the band were mounted 1m from the surface of the satellite how much longer would it have to be?

5. Imagine a rope long enough to go right round the Equator. How much longer would it have to be to encircle the Equator at a height of 1m from its surface? (The diameter of the Earth is 12 640 000m.)

6. A spaceship orbiting the Moon (diameter 3 480km) completes an orbit in eight hours. If its speed is 1 680km/h how far does it travel in one orbit? How far away from the Moon is its orbit?

Division puzzle

How good is your division? This puzzle will test your skill. In the number below all the digits are different.

38 125

The first digit, 3, can be divided exactly by 1.
The first two digits, 38, can be divided exactly by 2.
The first three digits, 381, can be divided exactly by 3.
The first four digits, 3 812, can be divided exactly by 4.
The first five digits, 38 125, can be divided exactly by 5.

Can you make up a five-digit number starting with 7 in which the digits work like this? (All the digits must be different.) Get a friend to check your answer.

Can you make up a six-digit or a seven-digit number like it?

There is one nine-digit number which works in the same way. Can you work out what it is?

Large number problems

Some problems involve numbers which are too big for a calculator to display. Most scientific calculators have a system called "scientific notation" for dealing with large numbers (pages 130-131) but here is a way of coping with them on simple calculators.

Adding and subtracting large numbers

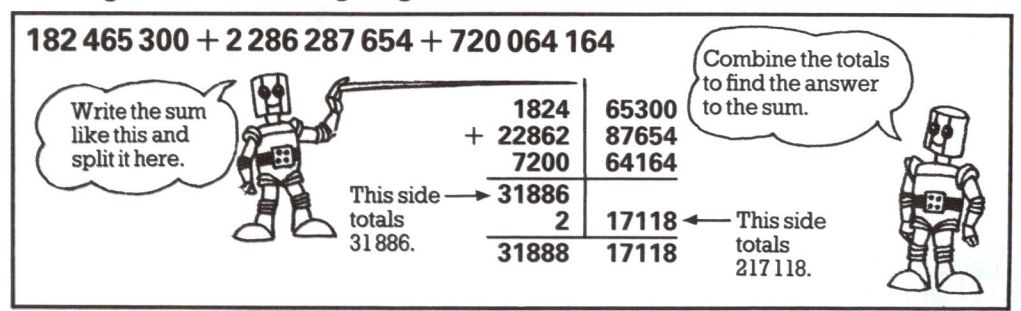

182 465 300 + 2 286 287 654 + 720 064 164

Write the sum like this and split it here.

```
      1824 | 65300
  + 22862  | 87654
      7200 | 64164
     31886
         2 | 17118
     31888 | 17118
```

This side totals 31 886.

This side totals 217 118.

Combine the totals to find the answer to the sum.

To add numbers which will not fit into the display, you can split the sum as shown above and add each side separately. Then combine the answers. Remember to carry a number from the righthand total across to the left if necessary.

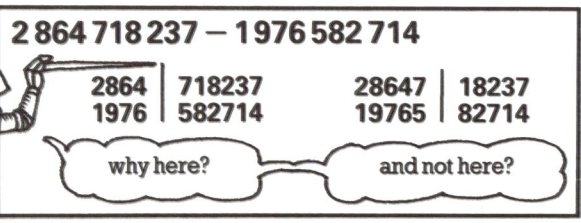

2 864 718 237 − 1 976 582 714

```
2864 | 718237        28647 | 18237
1976 | 582714        19765 | 82714
```

why here?   and not here?

You can use the same method to subtract numbers, but you have to be careful where you split the sum. Why would you split the above subtraction in the position shown on the left?

Examples to try

1.   5 266 834 710
    +   276 647 433
         27 164 311 803

2.   980 065 432
    −   735 917 141

3.   1 262 587 652 321 987 125
    +   921 766 412 005 286 421

How would you do this sum?

Multiplications and divisions

1   864 270 000 × 2 847

Divide the number on the left by 1 000 000 to convert to millions.

864.27 × 2847 =

`2460576.6`

Multiply by 1 000 000. The answer is 2 460 576 600 000 or about 2½ million million.

2   193 640 000 × 156 343 000

Divide each number by 1 000 000.

193.64 × 156.343 =

`30274.258`

Multiply twice by 1 000 000. The answer is 30 274 258 000 000 000 or about 30 thousand million million.

The trick for doing multiplications and divisions is to convert the numbers to millions and compensate for this afterwards. To convert numbers to millions you move the decimal point six places to the left (i.e. divide by 1 000 000).

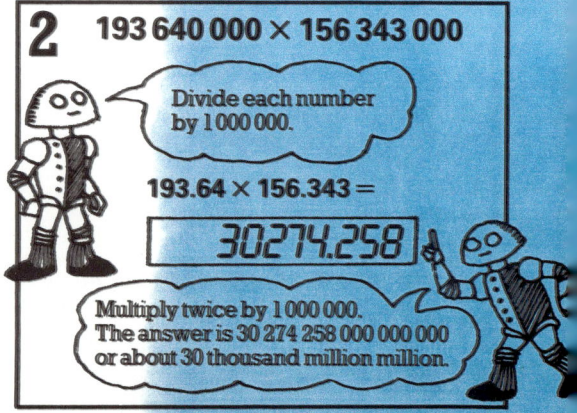

If both numbers are too big for the display you have to convert them both to millions and then multiply the answer by a million million to compensate (move the decimal point twelve places to the right).

3 86 427 000 000 ÷ 2 847

86 427 ÷ 2 847 = `30.357218`

Multiply by 1 000 000.
The answer is 30 357 218.

2 847 ÷ 86 427 000 000

2 847 ÷ 86 427 = `0.032941`

Divide by 1 000 000.
The answer is 0.000 000 03.

4 86 427 000 000 ÷ 288 090 000

86 427 ÷ 288.09 = `300`

This is the correct answer.

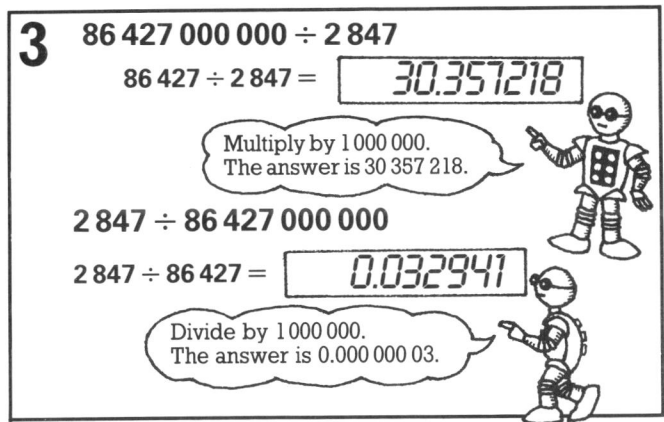

In divisions with big numbers the way you compensate the answer depends on which number in the sum you converted to millions. Decide what size of answer you would expect and then multiply or divide the calculator's answer accordingly.

If you convert both numbers in a division to millions the answer the calculator gives is correct.

Overflows

456 000 × 123 000

`560.8800E`

Some calculators have an "overflow check" and display an E if an answer is too large for the display. Try the example above. If your calculator displays an E you can find the right answer by moving the decimal point eight places to the right.

The E shows that the decimal point should be eight places to the right, so the correct answer is 56 088 000 000.

The overflow check prevents any further calculations with that number. However on some calculators you can release the overflow check by pressing "clear entry" and continue calculating. Remember to move the decimal point eight places to the right in your final answer.

Famous chessboard problem

The story is told of an Indian philosopher who helped his ruler in a time of great difficulty and was offered anything he wanted as a reward. The philosopher said he simply wanted 1 grain of rice on the first square of a chessboard, two grains on the second, four on the third, eight on the fourth and so on, for each square doubling the number on the previous square. The ruler laughed and was pleased that he did not have to part with any of his riches, but he soon stopped laughing. Use the constant function (2 × × =, or 2 × =) to find out why.

Sums inside sums

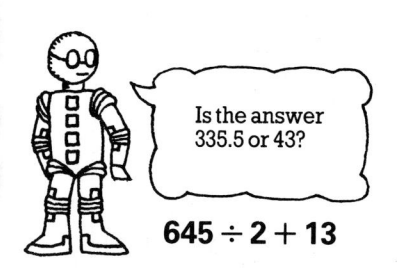

Is the answer 335.5 or 43?

645 ÷ 2 + 13

This calculation has two possible answers depending on whether you do the division or the addition first.

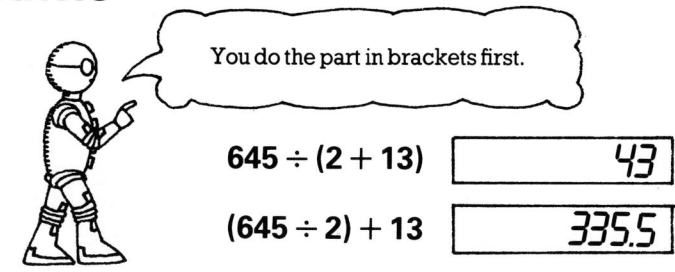

You do the part in brackets first.

645 ÷ (2 + 13) → 43

(645 ÷ 2) + 13 → 335.5

To avoid confusion in calculations with several parts, mathematicians use brackets to show which part should be done first. Many scientific calculators have brackets keys and automatically work out calculations in the correct order.

Examples

1. 41 − (32 − 21)
2. 117 + (39 ÷ 3)
3. (121 ÷ 11) − 10
4. 19 + (348 ÷ 16)
5. (200 − 135) ÷ 45
6. (47 − 7) ÷ (100 ÷ 5)

135 ÷ 15 − 6 + 5

→ 20
→ −2
→ 8

Try these calculations. If your calculator has brackets keys you can enter them as they are written. If it does not you need to decide the order. (You could use the memory to store the answer to one part while you do the other.)

Where would you put the brackets in this calculation to give the different answers shown above?

Calculating Easter

The date of Easter varies from year to year and is calculated from the phases of the Moon. (The Sun, Moon and Earth are lined up only once every 19 years. The intervening years are called the 19 phases of the Moon.) The calculation to find the date of Easter Saturday is shown below. It is quite long with lots of separate stages. The letters represent the answers and the remainders you get at each stage. To start the calculation you only need to know Y (the year). See if you can calculate when Easter will fall next year.*

Y ÷ 19 = Z, remainder A ← This division is to find out which phase the Moon is in.
Y ÷ 100 = B, remainder C ← This is to find out which century the year is in...
B ÷ 4 = D, remainder E ← ...and these two are to check for leap years.
(B + 8) ÷ 25 = F, remainder G
(B + 1 − F) ÷ 3 = H, remainder I
(19 × A) + (B + 15) − (D + H) = J
J ÷ 30 = K, remainder L
C ÷ 4 = M, remainder N
(2 × E) + (2 × M) + 32 − (L + N) = P
P ÷ 7 = Q, remainder R
A + (11 × L) + (22 × R) = S
S ÷ 451 = T, remainder U
(L + R + 114) − (7 × T) = V
V ÷ 31 = W, remainder X

W is the month, X the day and Y the year.

Y = 1990

Z = 104
B = 19
D = 4

REMAINDERS
A = 14
C = 90
E = 3

It may help to make a chart like this. If you are not sure how to work out the remainder from the calculator's answer, see Leftovers on page 139.

*You can check your answer in a diary.

Brackets inside brackets

1. $10 \div (10 \div (10 \div (10 \div (10 \div 10))))$

2. $10 \times [10 \div (10 \times \{10 \div [10 \times (10 \div 10)]\})]$

3. $$\cfrac{1}{\left(1+\left(\cfrac{1}{\left(1+\left(\cfrac{1}{(1+1)}\right)\right)}\right)\right)}$$

← Press the "close brackets" key five times before you press =.

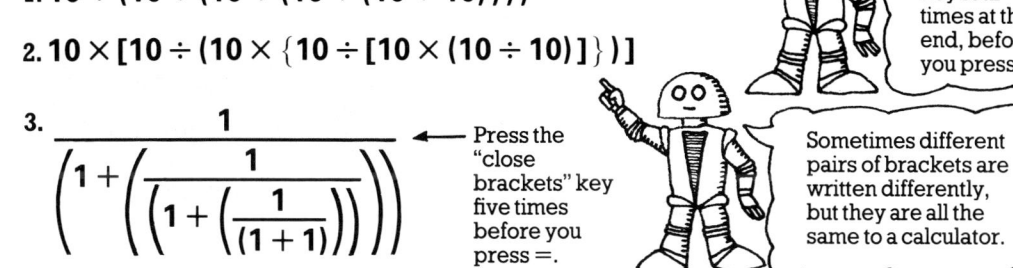

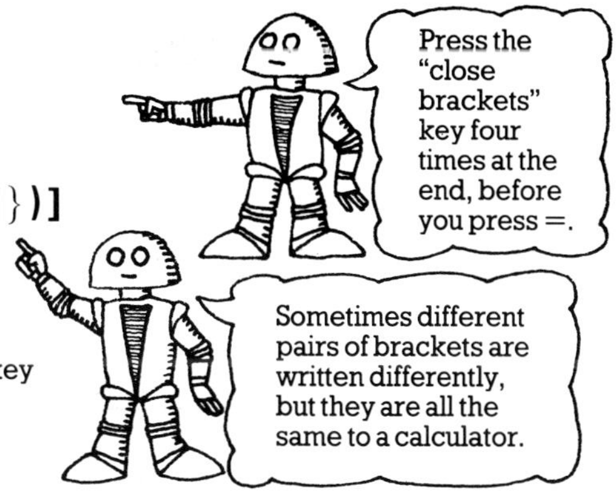

Press the "close brackets" key four times at the end, before you press =.

Sometimes different pairs of brackets are written differently, but they are all the same to a calculator.

Can you unravel these calculations? The rule for working out brackets inside brackets (nested brackets) is to do the sum in the innermost brackets first and work outwards. On a calculator with brackets keys you can enter the calculations in the order they are written, because when you press the "open brackets" key the calculator will not do the sum inside until you press the "close brackets" key.

Reciprocals*

A reciprocal is a number divided into 1, so $\frac{1}{3}$ is the reciprocal of 3 and $\frac{1}{5}$ is the reciprocal of 5. Reciprocals are simply fractions where the top part is 1 (called unit fractions) but they occur so frequently in calculations that many scientific calculators have a key for working them out. The reciprocal key divides any number you enter into 1 and expresses its reciprocal as a decimal. You can see how to use it on the right.

Reciprocal key

3 $\frac{1}{x}$ *0.3333333*

5 $\frac{1}{x}$ *0.2*

What is the reciprocal of 0.5?

What is the reciprocal of 0.1?

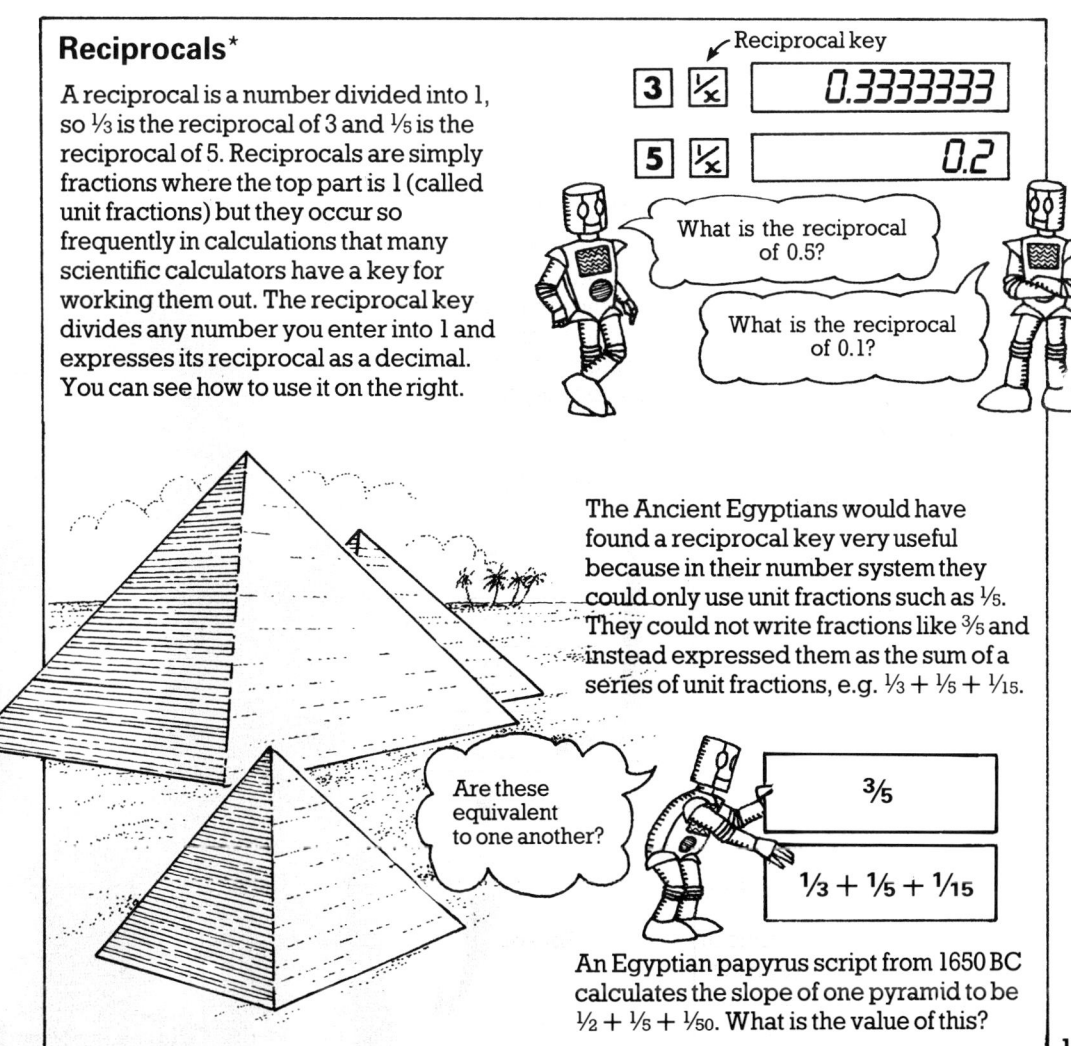

The Ancient Egyptians would have found a reciprocal key very useful because in their number system they could only use unit fractions such as $\frac{1}{5}$. They could not write fractions like $\frac{3}{5}$ and instead expressed them as the sum of a series of unit fractions, e.g. $\frac{1}{3} + \frac{1}{5} + \frac{1}{15}$.

Are these equivalent to one another?

$\frac{3}{5}$

$\frac{1}{3} + \frac{1}{5} + \frac{1}{15}$

An Egyptian papyrus script from 1650 BC calculates the slope of one pyramid to be $\frac{1}{2} + \frac{1}{5} + \frac{1}{50}$. What is the value of this?

*Go to pages 72-73 for more about reciprocals.

Power puzzles

The solution to the chessboard problem on page 29 involved multiplying 2 by 2 by 2 by 2 ... a total of 63 times. Mathematicians write this as 2^{63} (2 to the power of 63).

Powers on a calculator

| 2 | y^x | 7 | = | 128 |

7^2 2^{14} 2^{15} 2^{30}

On a scientific calculator you can use the power key, marked y^x, to multiply a number by itself.* For instance, the number of rice grains on the eighth square of the chessboard is 2^7 and you can work this out by pressing the keys shown above. Can you work out the other powers shown? If your calculator does not have a power key use the constant to multiply the numbers by themselves.

Problems with powers

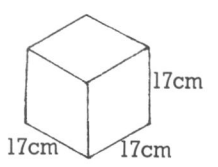

17cm, 17cm, 17cm

In mathematics there are many problems where you need to keep multiplying by the same number. The most common ones involve areas or volumes. For instance, the box above measures 17cm by 17cm by 17cm. Its volume is therefore 17^3ml (1 millilitre = 1 cubic centimetre).

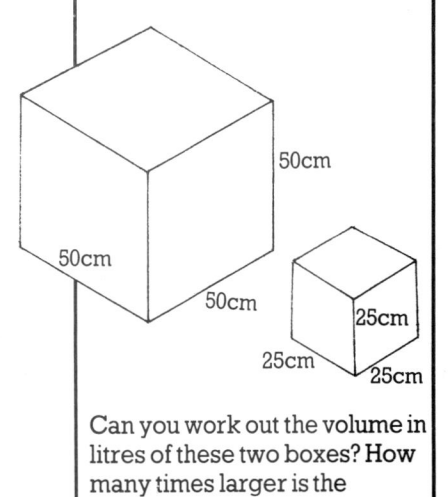

50cm, 50cm, 50cm

25cm, 25cm, 25cm

Can you work out the volume in litres of these two boxes? How many times larger is the volume of the bigger box?

Which is bigger?

Is 2^5 bigger than 5^2? Is 4^3 bigger than 3^4?

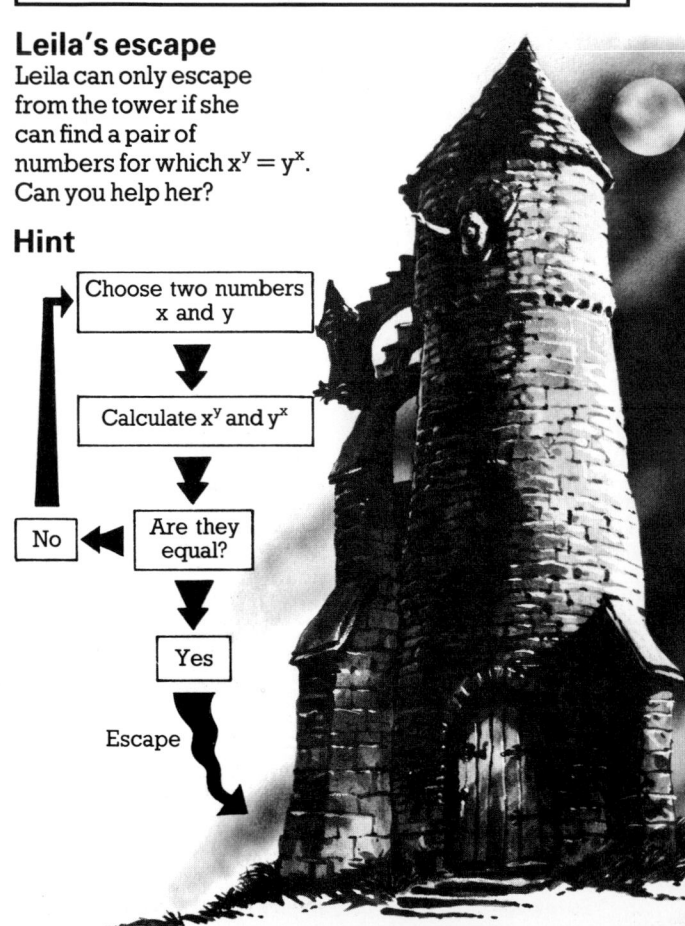

Leila's escape

Leila can only escape from the tower if she can find a pair of numbers for which $x^y = y^x$. Can you help her?

Hint

Choose two numbers x and y
↓
Calculate x^y and y^x
↓
Are they equal?
No → (back to Choose)
Yes ↓
Escape

*On some calculators the label is x^y.

Fractional powers

If your calculator has a power key you can work out "fractional" powers, such as $9^{1/2}$. The keys to press to work out $9^{1/2}$ are shown on the right. Can you work out the fractional powers and square roots shown below?

| 9 | y^x | 0 | . | 5 | = |

$9^{1/2}$ $\sqrt{9}$

$25^{1/2}$ $\sqrt{25}$

$64^{1/2}$ $\sqrt{64}$

The examples above show that the square root of a number is the same as the power "to a half". So $\sqrt{y}$ and $y^{1/2}$ are just two ways of writing the same thing. In fact all roots are fractional powers. To find the fourth root of, for example, 60 (that is the number which when multiplied by itself four times gives 60), you will need to calculate $60^{1/4}$.

| 6 | 0 | y^x | 0 | . | 2 | 5 | = | ← To check the answer multiply it by itself four times.

Matching powers and roots

64 is 2^6
$\sqrt{64}$ is 2^3

Are the robot's statements correct? You can check by working out the powers and square root. Then see if you can fill in the gaps in the table below.

$64 = 2^6$	$81 = 3^4$	$256 = 4^?$	$9 = 9^1$
$\sqrt{64} = 2^3$	$\sqrt{81} = 3^?$	$\sqrt{256} = 4^2$	$\sqrt{9} = 9^?$

1 Powers in fractions

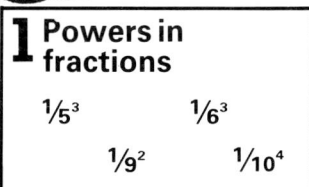

$1/5^3$ $1/6^3$

$1/9^2$ $1/10^4$

Can you work out the value of these fractions which involve powers? Remember you can use the reciprocal key to divide numbers into 1.

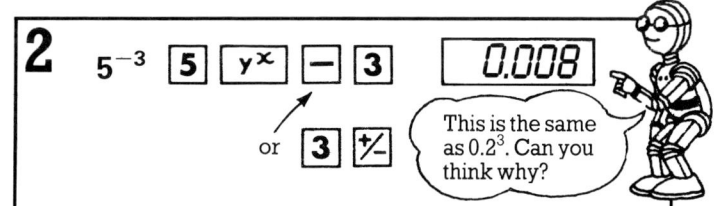

2

5^{-3} | 5 | y^x | – | 3 | $\boxed{0.008}$

or | 3 | +/– |

This is the same as 0.2^3. Can you think why?

Another way to write a fraction which involves a power is as a "negative power". For instance $1/5^3$ is the same as 5^{-3} and $1/6^3$ is the same as 6^{-3}. You can use the power key on a calculator to work out negative powers, entering them as shown above.*

FIX key

$150 \div 7$

Calculate this to two, three and four decimal places.

Divide $78 by 11 and give the answer in dollars and cents.

Many scientific calculators have a key marked FIX which makes the calculator round off answers to the number of decimal places you require. To use it you press FIX and then the number of places you want, before starting a calculation. Try the example above.

The FIX key is useful for doing calculations with money. Most currencies are based on units of 100 (a dollar is 100 cents, a UK pound is 100 pence, a French franc is 100 centimes), so you need to round answers to two decimal places.

*See page 130 if your calculator gives the answers in scientific notation.

More about large numbers

To display very large numbers scientific calculators use a system called Standard Form or scientific notation. It uses the fact that a number like 300 000 000 000 000 (three hundred million million) can also be written as 3 × 100 000 000 000 000, or to avoid writing out the zeros, 3×10^{14}.

$$6\,829\,700 = 6.8297 \times 1\,000\,000 = 6.8297 \times 10^6$$

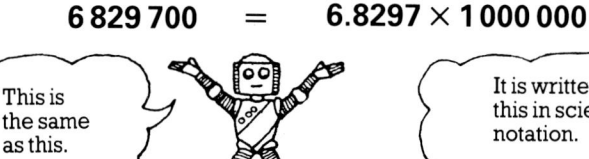

This is the same as this.

It is written like this in scientific notation.

In scientific notation a number is always expressed as a number between 1 and 10, multiplied by a power of ten as shown in the example above.

Number between 1 and 10 × **Power of ten**

Converting numbers

Can you convert these numbers to scientific notation?

In these pairs of numbers which is bigger?

1. 743 800 000 000
2. 9 230 000 000
3. 802 000 000 000
4. 45 320 000 000

5. 1.49×10^8 or 153 000 000
6. 587 000 000 or 4.17×10^{10}
7. 9.3×10^8 or 3.8×10^9
8. 1.9×10^6 or 9.1×10^5

Very small numbers

Scientific notation is also used to display very small numbers. These often occur in microbiology. For instance the smallest living cells are a kind of bacteria which have a diameter of 0.000 025mm. Blood cells have a diameter of 0.000 75mm.

$$0.000\,75 = 7.5 \times {}^1/_{10\,000} = 7.5 \times 10^{-4}$$

This is the same as this.

It is written like this in scientific notation.

When very small numbers are written in scientific notation the power of ten is negative, because multiplying by a negative power is the same as dividing by a power (see page 129).

Small number puzzle

Can you arrange these in order of size starting with the smallest? (The sizes given are diameters.)

Pneumonia bacteria = 0.000 1mm

Paramecium protozoa = 0.2mm (single-celled organisms)

'Flu virus = 5×10^{-5}mm

Mumps virus = 0.000 225mm

Molecule of egg white protein = 0.000 01mm

Atom of hydrogen = 2×10^{-7}mm

Pin prick = 10^{-1}mm

Scientific notation on a calculator

987 654 × 456 789

Display: `4.5114948 11`

That means 4.5114948×10^{11} = 451 149 480 000

If you do the above calculation on a scientific calculator it will show the answer in scientific notation. The number at the right of the display is the power of 10 and it is called the exponent.

Puzzles

If you have a scientific calculator try these problems and puzzles.

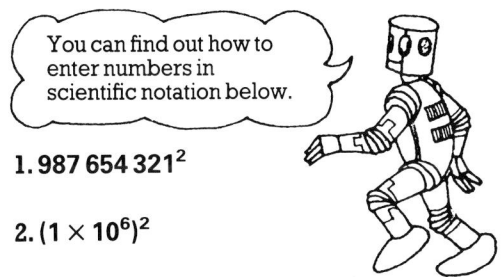

You can find out how to enter numbers in scientific notation below.

1. $987\,654\,321^2$

2. $(1 \times 10^6)^2$

3. $(2.6 \times 10^{16}) \div (1.3 \times 10^{14})$

4. How many heartbeats are there in a lifetime? (The average pulse rate is 70 beats a minute.)

5. Light travels at the speed of 2.998×10^5 km/sec. How far is a light year (the distance travelled by a ray of light in one year)?

6. 7.5×10^{-4} mm is the diameter of a blood cell. How many times larger is this than a bacteria cell, which is 2.5×10^{-5} mm in diameter?

7. How many times larger is a 'flu virus than a hydrogen atom? (Their sizes are given in the Small number puzzle on the left.)

8. Fleas are very good at jumping. A flea's take-off power is 2×10^7 ergs per gram per second. An adult person's take-off power is only 5×10^5 ergs per gram per second. If an adult can jump 1.75 metres, how high could a flea of the same weight jump?

An erg is a measurement of energy.

Entering numbers in scientific notation

2.6×10^{16}

[2] [.] [6] [EXP] [1] [6] ← Exponent key

These are the keys to press to enter a number in scientific notation. When you press EXP the calculator expects the next number to be a power of ten.

7.5×10^{-4}

[7] [.] [5] [EXP] [4] [+/−]

Very small numbers will have a negative power. To enter a negative power press the change sign key after the power, as shown here.

SCI key

Many scientific calculators have a key labelled SCI which makes the calculator display answers in scientific notation. However you need to set the number of significant figures the answer should have. To do this you press SCI, then the number of significant figures you want, before starting a calculation. If your calculator has a SCI key try these.

1. $267.45 \div 17.862$
2. $114 \div 21.68$
3. $29\,764 \times 3\,968$
4. $(2.96 \times 10^3) \div (8.914 \times 10^5)$

Give the answers to three significant figures.

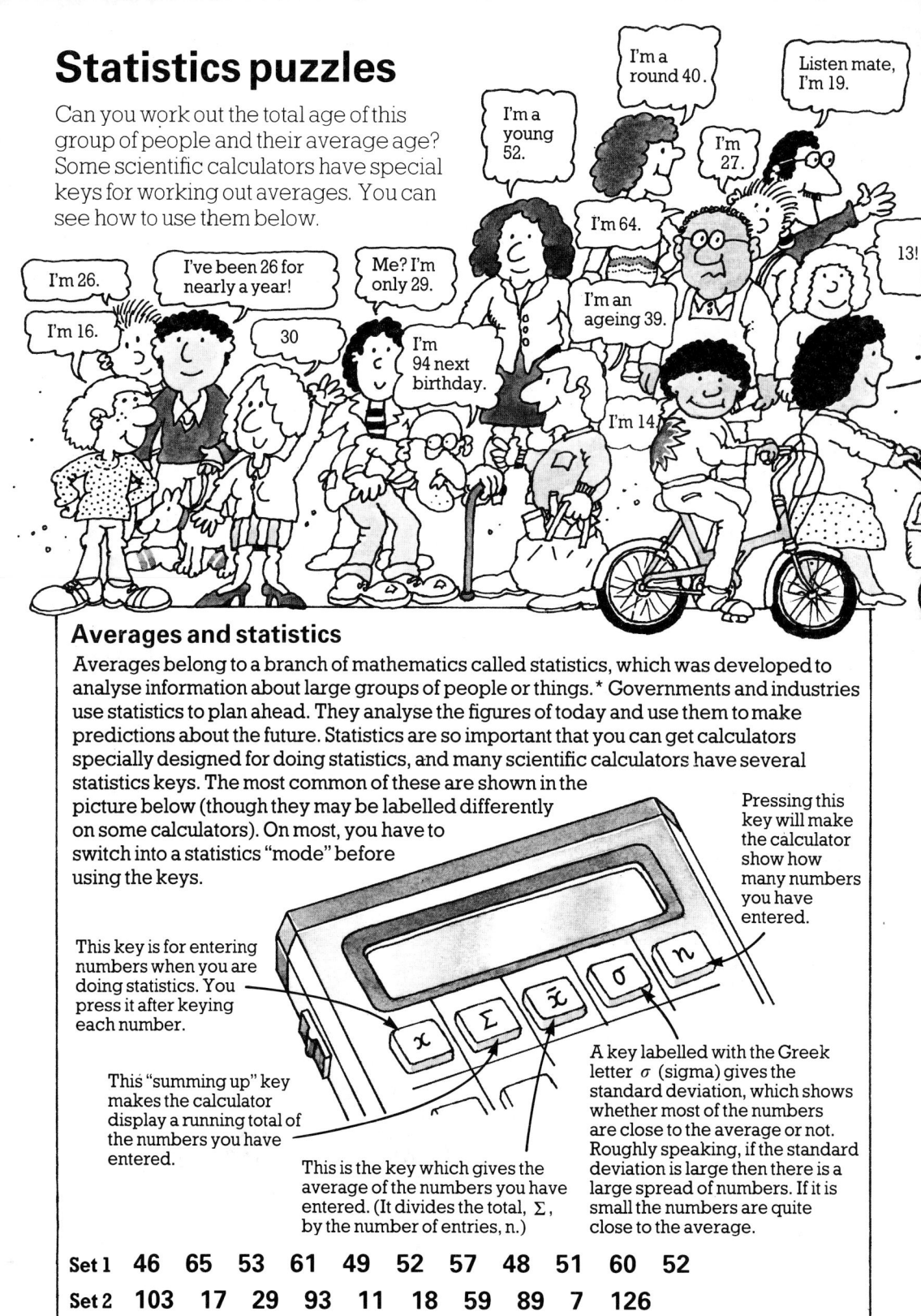

Statistics puzzles

Can you work out the total age of this group of people and their average age? Some scientific calculators have special keys for working out averages. You can see how to use them below.

Speech bubbles:
- I'm a round 40.
- Listen mate, I'm 19.
- I'm a young 52.
- I'm 27.
- I'm 64.
- 13!
- I'm 26.
- I've been 26 for nearly a year!
- Me? I'm only 29.
- I'm an ageing 39.
- I'm 16.
- 30
- I'm 94 next birthday.
- I'm 14.

Averages and statistics

Averages belong to a branch of mathematics called statistics, which was developed to analyse information about large groups of people or things.* Governments and industries use statistics to plan ahead. They analyse the figures of today and use them to make predictions about the future. Statistics are so important that you can get calculators specially designed for doing statistics, and many scientific calculators have several statistics keys. The most common of these are shown in the picture below (though they may be labelled differently on some calculators). On most, you have to switch into a statistics "mode" before using the keys.

Pressing this key will make the calculator show how many numbers you have entered.

This key is for entering numbers when you are doing statistics. You press it after keying each number.

This "summing up" key makes the calculator display a running total of the numbers you have entered.

This is the key which gives the average of the numbers you have entered. (It divides the total, Σ, by the number of entries, n.)

A key labelled with the Greek letter σ (sigma) gives the standard deviation, which shows whether most of the numbers are close to the average or not. Roughly speaking, if the standard deviation is large then there is a large spread of numbers. If it is small the numbers are quite close to the average.

| Set 1 | 46 | 65 | 53 | 61 | 49 | 52 | 57 | 48 | 51 | 60 | 52 |
| Set 2 | 103 | 17 | 29 | 93 | 11 | 18 | 59 | 89 | 7 | 126 |

If your calculator has statistics keys see if you can work out the sum, average and standard deviation for the two sets of numbers above.

*See also pages 30-31 and 86-87.

More statistics questions

1. There are more people aged under 30 in this sample than over 30. Why is the average age not under 30?
2. Assuming this sample of people is representative of the population as a whole, what percentage of people are over 60 years old and what percentage are under 20?
3. Judging by this sample what proportion of TV programmes should be for children under ten?

Decimal maze

For a change from statistics, see if you can work out how to get through this maze. You start with a 1 in your calculator display and as you travel along each path you must multiply the number in the display by the number you pass. You must finish with a 5 in the display and you may only travel along each path once, but it can be in any direction.

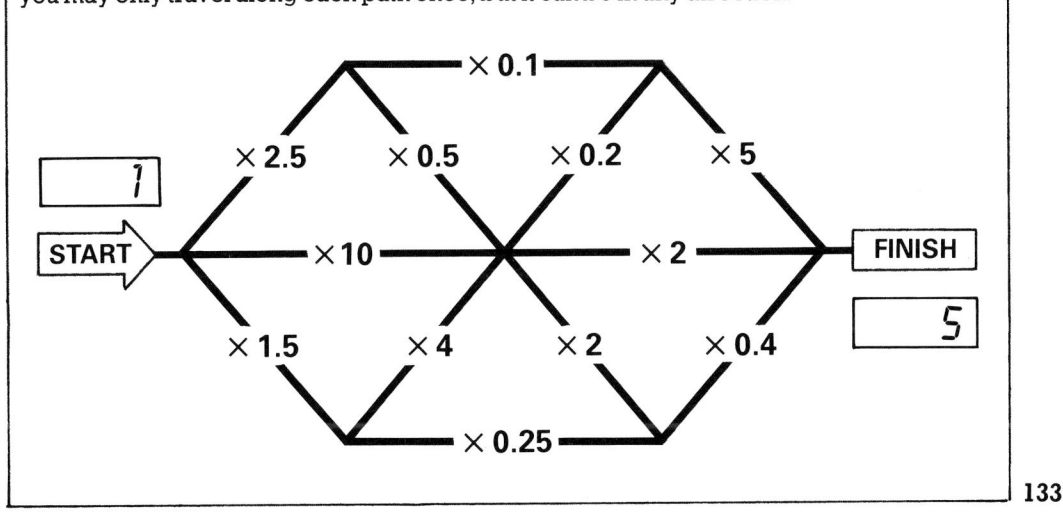

Likely or unlikely?

A branch of mathematics very closely related to statistics is called probability.* People making decisions about the future may need to know how likely an event is to happen. For example, before deciding to build a flood barrier they need to know what the probability is of the river flooding. There are mathematical ways of working out probability, but they can only ever provide a guideline. The probability of getting tails when you toss a coin, for instance, is 50-50 which is the same as 1 in 2 or ½, but a coin does not always come up tails exactly half the times you toss it – you might get five tails in a row. However if you tossed the coin a hundred times you could expect to get tails between 45 and 55 times (though you could not be certain).

What is the probability of choosing an ace from a pack of playing cards?

There are four aces in a pack of 52 cards so the probability of getting an ace is $4/52$.

$4/52 = 0.076\,923$, that is **0.08**

This is quite a small probability. What is the probability of choosing a diamond from the pack?

Probability scale

Probability may be expressed as a fraction, or as a decimal or a percentage. If an event is almost certain (for instance, the sun rising tomorrow) its probability is almost 1 or 100%. If an event is very unlikely (for instance, that you will get measles tomorrow) the probability is near to 0 or 0%. Where would the events below go on the probability scale shown on the right?

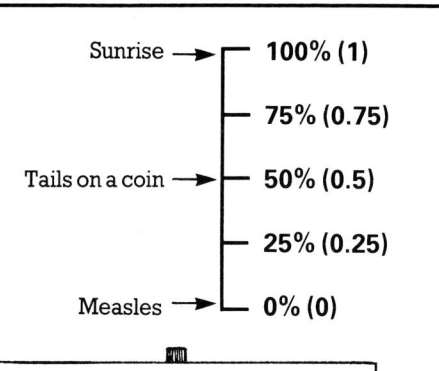

Sunrise → 100% (1)
— 75% (0.75)
Tails on a coin → 50% (0.5)
— 25% (0.25)
Measles → 0% (0)

Getting a six on a dice.

Having a birthday in October.

Having a birthday between January 1st and August 31st.

Multiplying probabilities

$4/52 \times 13/52 = 0.019$

This is the same as $1/52$.

If you are considering the likelihood of two events happening at once you need to multiply their separate probabilities. For instance, the probability of picking an ace from a pack of cards is $4/52$. To pick a diamond the probability is $13/52$. For both things to happen (an ace of diamonds) the probability is $4/52 \times 13/52$.

*See also pages 42-44.

Aces puzzle

If you deal four cards from a pack what is the probability of getting four aces?

The probability of the first card being an ace is $4/52$. If that was successful there are now 51 cards and three aces left, so the chances of getting another ace are $3/51$. If the second card is an ace too, there are 50 cards and two aces left so the probability of the third card being an ace is $2/50$. The probability of the fourth ace coming up next is $1/49$. The probability of getting all four aces is therefore:

$$4/52 \times 3/51 \times 2/50 \times 1/49$$

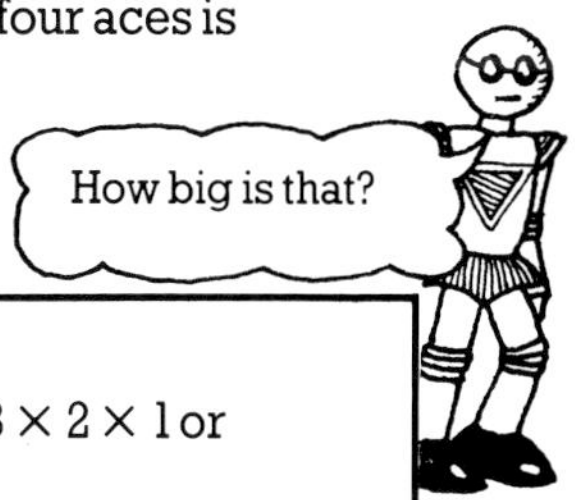

How big is that?

Factorials

Multiplying probabilities often creates sequences of numbers like $4 \times 3 \times 2 \times 1$ or $6 \times 5 \times 4 \times 3 \times 2 \times 1$. These are called factorials and are written 4! or 6!.

[4] [x!] ← Factorial key

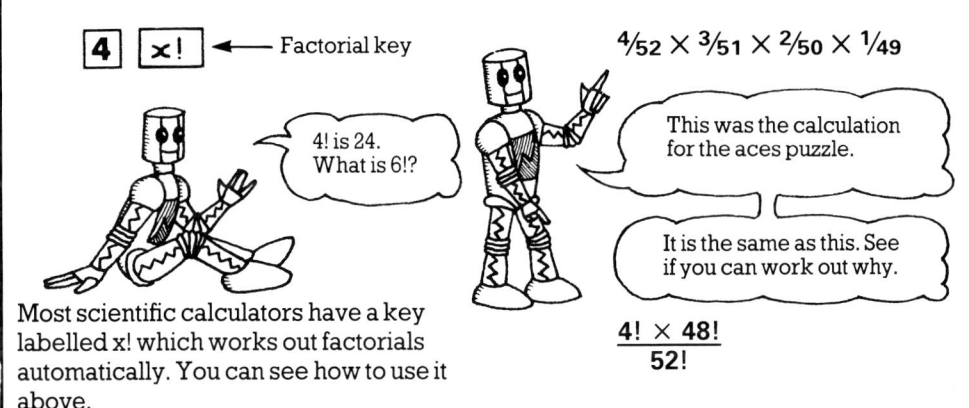

4! is 24. What is 6!?

$$4/52 \times 3/51 \times 2/50 \times 1/49$$

This was the calculation for the aces puzzle.

It is the same as this. See if you can work out why.

Most scientific calculators have a key labelled x! which works out factorials automatically. You can see how to use it above.

$$\frac{4! \times 48!}{52!}$$

Adding probabilities

If an outcome depends on one or another event happening you need to add their separate probabilities. For instance, this spaceship landing could be disastrous if the surface of the planet is too soft or the spaceship's speed on landing is too fast, or if the spaceship overheats on entering the planet's atmosphere or its engines fail.

Puzzle

Given the probabilities listed below, can you work out how safe the landing is? If the danger of mishap is more than 1 in 4 the landing must be aborted.

1. Scientists have predicted that the possibility of the planet having a very soft surface is 1 in 5.
2. The speed of the landing can only be predicted to an accuracy of 1 in 35.
3. There is a 1 in 60 chance of the spaceship overheating but the chances of engine failure are only 1 in 200.

135

Puzzles with triangles

The puzzles on these two pages involve right-angled triangles, so you can use the rules of trigonometry to solve them. You can find out about trigonometry below. (You will need a calculator with keys labelled tan, sin and cos.)

Mine rescue
Some miners are trapped by an explosion 80 metres along a tunnel which is 130 metres below the pithead. An emergency air-passage is to be drilled down to the miners. At what angle should it be drilled?

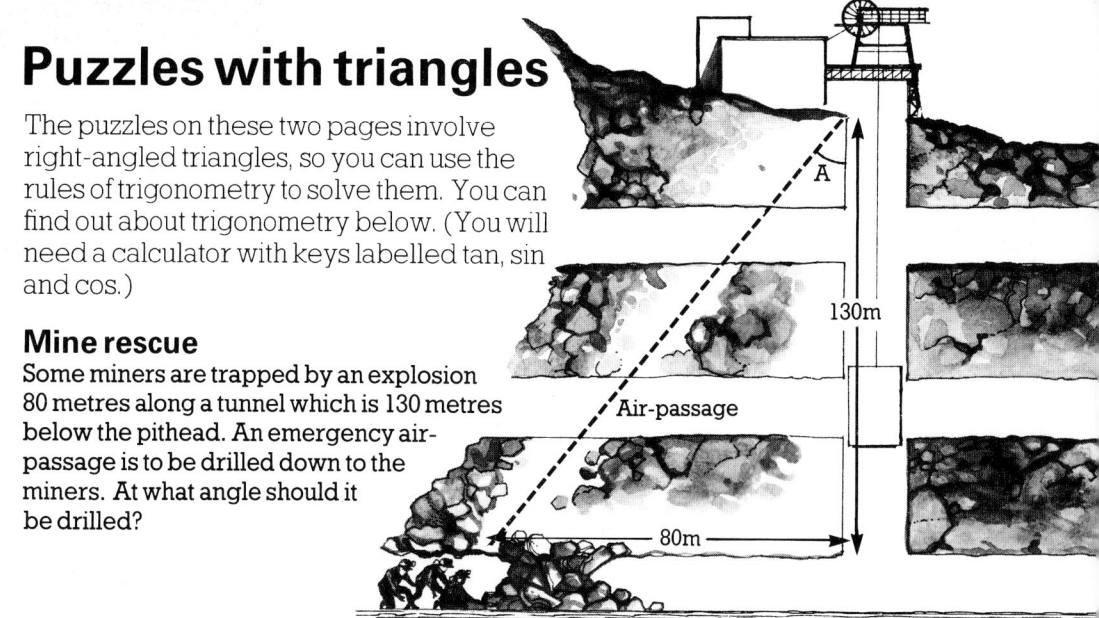

Trigonometry

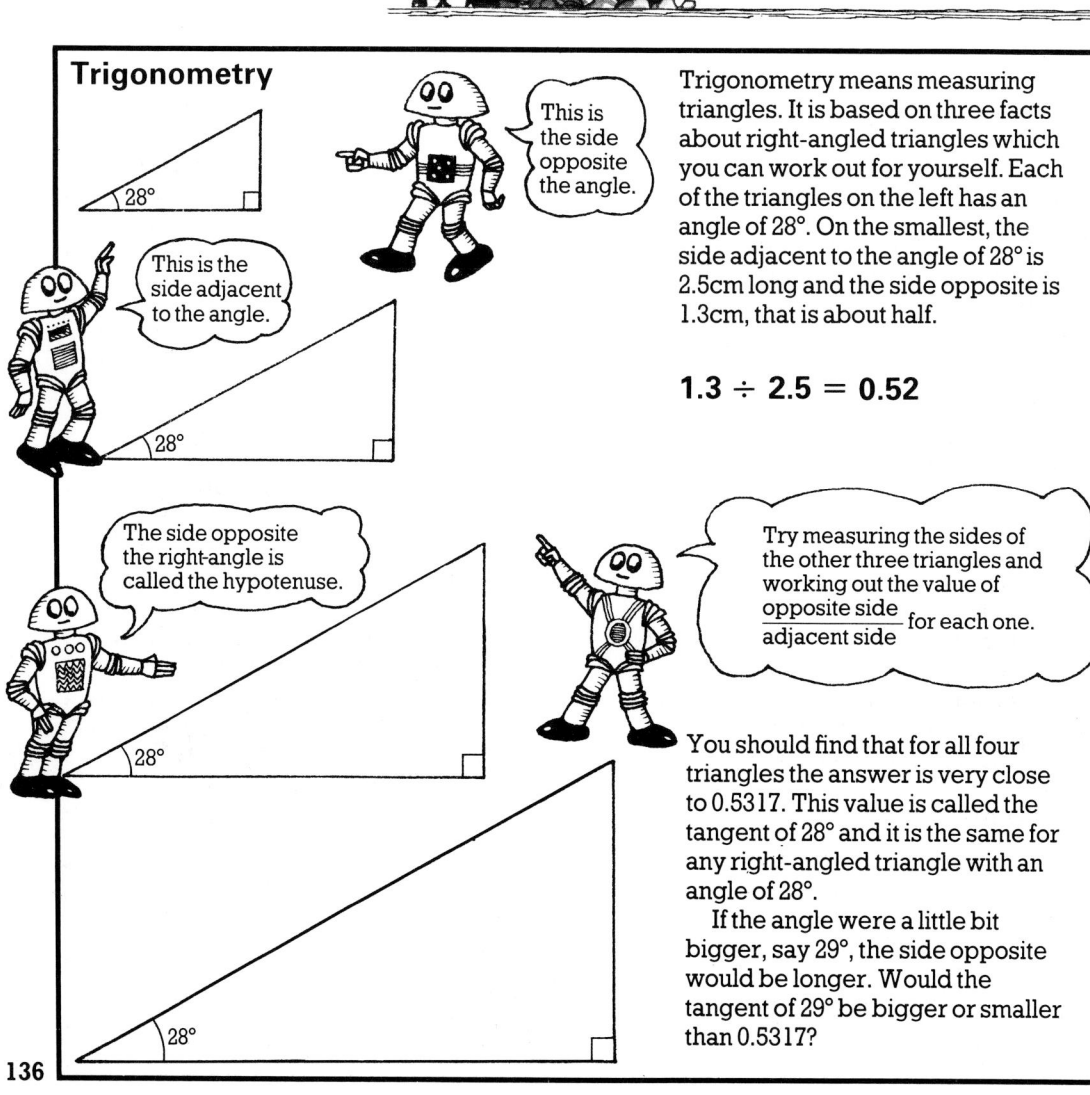

This is the side opposite the angle.

This is the side adjacent to the angle.

The side opposite the right-angle is called the hypotenuse.

Trigonometry means measuring triangles. It is based on three facts about right-angled triangles which you can work out for yourself. Each of the triangles on the left has an angle of 28°. On the smallest, the side adjacent to the angle of 28° is 2.5cm long and the side opposite is 1.3cm, that is about half.

1.3 ÷ 2.5 = 0.52

Try measuring the sides of the other three triangles and working out the value of $\frac{\text{opposite side}}{\text{adjacent side}}$ for each one.

You should find that for all four triangles the answer is very close to 0.5317. This value is called the tangent of 28° and it is the same for any right-angled triangle with an angle of 28°.

If the angle were a little bit bigger, say 29°, the side opposite would be longer. Would the tangent of 29° be bigger or smaller than 0.5317?

Tower puzzle
Standing 20 metres from a tower the angle to the top is 52°. How tall is the tower?

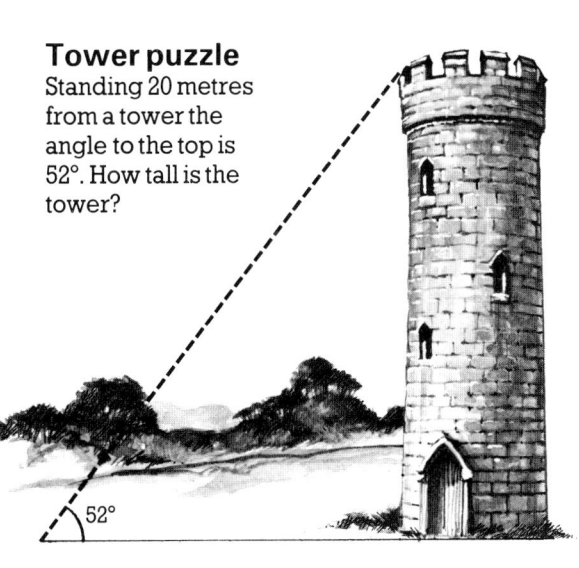

Comet alarm
An unknown comet is fast approaching Earth. It is estimated to be 180 000km away, travelling at an angle of only 2°. Remembering that the radius of the Earth is about 6 400km, will the comet collide with Earth?

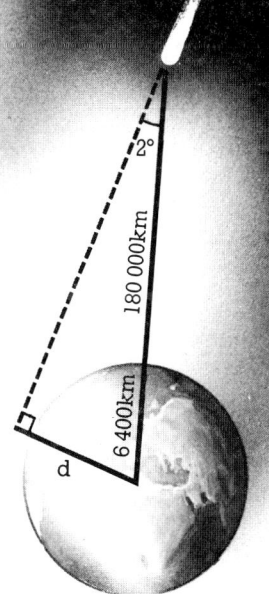

Sines and cosines
You could also calculate $\frac{\text{opposite side}}{\text{hypotenuse}}$ or $\frac{\text{adjacent side}}{\text{hypotenuse}}$ for the four triangles. Each of these gives a constant value too. The values, or ratios*, are called the sine and cosine of the angle of 28°.

$\frac{\text{opposite side}}{\text{hypotenuse}} = \text{sine}$

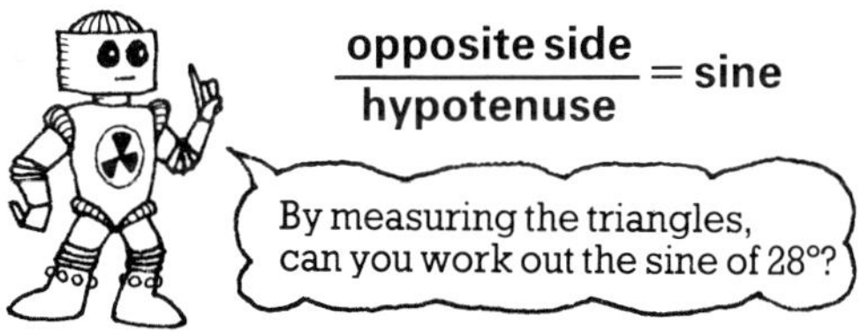

By measuring the triangles, can you work out the sine of 28°?

$\frac{\text{adjacent side}}{\text{hypotenuse}} = \text{cosine}$

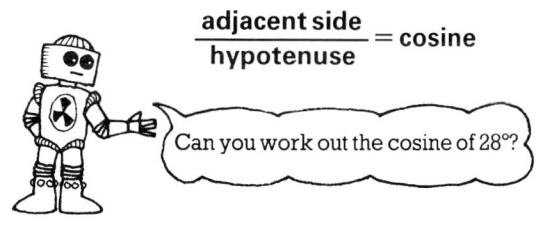

Can you work out the cosine of 28°?

Using trigonometry
The three ratios, sine, cosine and tangent are useful because they enable people like surveyors, astronauts and astronomers to calculate distances and angles which are difficult or even impossible to measure. Scientific calculators have keys which automatically give you the sine, cosine or tangent of any angle. If you have a scientific calculator check this example:

`3` `2` `tan`   `0.6248693`

A scientific calculator will also reverse the process and give you the angle if you enter the sine, cosine or tangent. To do this you need to press the inverse key before the sin, cos or tan key. The example below shows the keys to press to find the angle whose tangent is 0.5317.

`0` `.` `5` `3` `1` `7` `INV` `tan`

`27.999578`

Trig puzzles

1. Can you work out the lengths of sides b and c in this triangle?

2. Can you work out the sizes of the angles x and y in this triangle?

*A ratio is a way of expressing the relationship between two numbers by dividing one into the other.

Puzzle answers

Page 99
1. You can make 12 with only five presses (11 + 1 =).
2. The figure 1 occurs 21 times in the numbers 1 to 100.

Page 100
Keyboard puzzles
1. Adding all the numbers from 1 to 20 gives a total of 210. If you did it in less than 25 seconds you were very quick.
   Here is an even quicker way to do it. The numbers between 1 and 20 can be divided into pairs, each of which adds up to 21 (1 + 20, 2 + 19, and so on). There are ten pairs so the answer is 10 × 21 = 210.
2. The following calculation will make the calculator display 100 with only ten presses: 37 + 73 − 7 − 3 =.
3. This is the calculation to make 1 001 with nine presses: 72 × 7 × 2 − 7 =.
4. The whole numbers which will divide exactly into 1 001 are 7, 11, 13, 77, 91, 143. They are called the factors of 1 001.
5. The operation keys were − and × (87 − 19 × 31 = 2 108).
6. The complete calculation is 48 × 73 = 3 504.
7. The two possible calculations are 93 × 86 = 7 998 and 83 × 86 = 7 138.

Page 101
Subtraction puzzles
The answer is always 111 if you subtract the columns of digits because each number on the keyboard is 1 larger than the one on its left.
   Subtracting the rows of digits you always get the answer 333 because each number is 3 more than the one below.

Pairs
When you subtract the two numbers the difference is always 27. This is because for each pair of keys the difference between the digits is always 3. When you combine the two digits e.g. 4 and 1 to make 41 and 14, the difference between the "tens" is always 30 and the difference between the units is always −3, so the combined difference is 27.
   The numbers (e.g. 41 and 14) always add up to a multiple of 11 because the sum of the tens is always the same as the sum of the units (4 + 1 = 5 and 1 + 4 = 5). In numbers below 100, if the tens and units are the same the number is divisible by 11, because 11 is one ten and one unit.

Neighbours
The five numbers which cannot be made are 8, 12, 14, 16 and 19. Using diagonal neighbours you can make 8 (5 + 3), 12 (6 × 2) and 14 (8 + 6) but 16 and 19 still cannot be made.

Page 102
Space highway
To move along the highway the numbers you need to add and subtract are: − 600; + 4 000; + 1; + 900; + 20; − 9 000; + 2; − 1 000; + 9; + 80; + 9; + 600; + 7; + 70; − 9; − 90; + 5 000; + 1.

Page 103
What is it worth?
1. The value of 4 in 36 417 is 400. If you need to check, subtract 400 from 36 417 on a calculator. The value of 4 in 29 149 is 40.
The value of 4 in 42 613 is 40 000.
2. The difference between 4 762 and 4 062 is 700.
3. The difference between 472 and 402 is 70.

Matching operations
The operations which are equivalent are:
A, C and I; B, E and J; D and G; F and H.

Page 104
Which is closest?
1. Answer A is closest.
2. Answer C is closest.
3. Answer C is closest.
One way to estimate answers is to round off the numbers in the sum to the nearest ten (or hundred or thousand), so you can do the sum in your head. For instance, in the first example 97 × 49 is about 100 × 50.

Find the mistakes
The third calculation is correct. The correct versions of the other calculations are shown below.
1. 987 × 3 = 2 961, so key 6 was pressed instead of key 3.
2. 1 629 + 1 332 = 2 961, so the − key was pressed instead of the + key.
4. 423 × 7 = 2 961, so 432 was pressed instead of 423.

Down to zero
Any four-figure number can be reduced to zero in four steps using the method shown in the example, i.e.
1. Subtract the last two figures.
2. Divide by the first two figures. (This always gives an answer of 100.)
3. Subtract 50.
4. Subtract 50 again.

Race to 1
This is the calculation to reduce 28 to 1 in three steps, using the key number 3:
28 + 3 = 31 + 3 = 34 − 33 = 1
Here are suggestions for reducing the other numbers:
55 − 66 + 6 + 6 = 1
40 + 5 + 5 + 5 ÷ 55 = 1
27 − 7 × 7 + 7 + 7 − 77 ÷ 77 = 1

Page 105
Follow ons
1. 15 + 16 + 17 + 18 = 66; 2. 34 × 35 = 1 190;
3. 7 × 8 × 9 = 504.

Reversing puzzle
In each case the figure you need to subtract is always a combination of numbers which are multiples of 9, for instance 272727, or 454545. The multiple of 9 depends on the difference between the starting digits. For example, the difference between the digits 8 and 3 is 5, 5 × 9 is 45 and the number you need to subtract from 838383 is 454545.

Page 106
Division trick
The trick could be: "Enter any four-digit number. Repeat it to make an eight-digit number. Divide by 73, then by 137".

Leftovers
To work out the remainder from the calculator's answer, you need to subtract the part of the answer before the decimal point. Then multiply the part after the point by the number you divided by, as shown below.

$$1001 \div 5 = 200.2$$
$$- 200 = 0.2$$
$$\times 5 = 1 \leftarrow \text{This is the remainder.}$$

Matching puzzle
17 ÷ 2 = 8.5; 107 ÷ 10 = 10.7; 100 ÷ 8 = 12.5; 37 ÷ 4 = 9.25; 8 ÷ 16 = 0.5; 12 ÷ 48 = 0.25; 54 ÷ 12 = 4.5; 10 ÷ 8 = 1.25.

Page 107
More matching puzzles
30 ÷ 7 = 4.285 714 2; 26 ÷ 11 = 2.363 636 3; 10 ÷ 3 = 3.333 333 3; 17 ÷ 4 = 4.25; 5 ÷ 18 = 0.277 777 7; 9 ÷ 16 = 0.562 5.

Which is bigger?
These are the answers to the calculations and the cards they belong to: 7 ÷ 3 = 2.333 333 3 – card B; 49 ÷ 19 = 2.578 947 3 – card C; 440 ÷ 200 = 2.2 – card A.

Page 108
Sizing up fractions
Starting with the smallest the order of the fractions is $1/5$ (= 0.2), $104/498$ (= 0.208 835 3), $225/1042$ (= 0.215 930 9), $18/79$ (= 0.227 848 1), $7/30$ (= 0.233 333 3), $41/170$ (= 0.241 176 4).

Page 109
Lost fraction puzzle
The robot's calculation was 3 ÷ 13 = 0.230 769 2.

Tops and bottoms
$6/11$ = 0.545 454 5 which is more than 0.5.

Finding patterns
The first twelve digits for $1/7$ are 0.142 857 142 857.
 The complete sixteen-digit pattern for $1/17$ is 0.058 823 529 411 764 7, and for $6/17$ it is 0.352 941 176 470 588 2.

Fractions made easy
$3/10 \times 2½ = 0.75$; $½ \div ¼ = 2$; $3/5 \times 4/7 = 0.342\ 857\ 1$; $5\frac{1}{8} - 3\frac{3}{4} = 1.375$; $7/8 + 3/4 = 1.625$; $½ \times ½ = 0.25$.

Page 110
Being too accurate
1. A light year is nearly 6 million million (6 000 000 000 000) miles or more than 9 million million (9 000 000 000 000) kilometres.
2. There are just over 30 million (30 000 000) cats in the USA.
3. The population of London is about 7 million.
4. The shortest street in Britain is about 17½ metres long.
 Rounding off the numbers in the statements simplifies them and helps you get a feeling for how big they are.

Puzzles
1. The modern train is nearly nine times faster than the 1829 record holder.
2. It would take the train just over three days to complete the journey.
3. A million hours is 41 666.66 days, which is about 114 years, so unless you are very, very old the answer is no.

Three-figure accuracy
Rounded to three significant figures the first four numbers are: **1.** 0.006 38; **2.** 0.062 9; **3.** 264; **4.** 781 000.
In example 5 the number 0.199 999 9 cannot be rounded to three significant figures. It becomes 0.200, that is 0.2.

Page 111
Space puzzles
1. The spaceship's journey would take 240 000 seconds, which is 4 000 minutes or 66.666 666 hours. So your answer could be "nearly 67 hours". You may have calculated that it is 2.777 777 7 days, so your answer could be "just over 2¾ days" or "just over 2 days and 18 hours".
2. The spaceship will reach Mars after about 58 days.
3. The Sun is approximately 400 times bigger than the Moon. It appears to be the same size because it is about 400 times further away.

Page 112
Repeats
1. The answers are 11, 111, 1 111, 11 111.
To get an idea of why this pattern occurs look at one of the calculations, for instance 1 234 × 9 + 5.
 1 234 × 9 is equivalent to
 1 000 × 9 = 9 000
 + 200 × 9 = 1 800
 + 30 × 9 = 270
 + 4 × 9 = 36

In this addition each column, except the units, totals 10. Adding 5 to the units and then carrying the 1s makes each column total 11. The other calculations work in the same way.
2. The pattern is 2 002, 3 003, 4 004, and so on. The clue to this pattern is that 143 × 7 = 1 001. In fact you could have stored 1 001 in the memory since both 143 and × 7 appear in every sum.
3. The pattern of answers is 9, 98, 987, 9 876, and so on. To investigate how it occurs, take one of the calculations, e.g. 123 × 8 + 3, and follow the same steps as in the answer to example 1.

139

Page 113
Conversions
85km is about 53 miles; 126km is about 79 miles; 100km is about 63 miles; 32km is 20 miles; 50km is about 31 miles; 10km is about 6 miles; 270km is about 169 miles; 115km is about 72 miles.

Order of calculations
$\frac{285 + 117}{264 - 68} = 2.0510204$

Who wins?
The woman takes just over 90 seconds to finish the race. That is 1 minute and 30 seconds, so she just beats the horse.

Page 114
Upside-down planet puzzle
To decode the answers turn the calculator upside-down and read the figures as letters. Starting at Earth the journey is via Sol, Lilosi, Zigo, then Isis, Lesbos and Sibel to Gogol. The final message at Gogol is "seize Sol".

Page 115
Investigating constants
5. Ten presses brings you closest to 100.
6. The number in the display gets bigger because you are subtracting a negative amount each time.
7. You need to press 9 + + = to display the nine times table.

Plant puzzle
After 15 days the plant would be 32 768cm high, that is over 300 metres.

Page 116
Percentage puzzles
If 20% of the population of Brigg emigrate the number of people leaving the island is 49 800 and the number left behind is 199 200. You can check your answers by adding them together to see if the total is the same as the original population.

20% of 4 000 = 800; 20% of 12.5 = 2.5; 90% of 8.9 = 8.01; 50% of 500 = 250.

Percentage chart
This is the completed chart:

50%	50/100	½	0.5
25%	25/100	¼	0.25
10%	10/100	1/10	0.1
33⅓%	33.33/100	⅓	0.3333
15%	15/100	3/20	0.15
66⅔%	66.67/100	⅔	0.6667

Percentages without a % key
15% of 30 = 4.5; 50% of 50 = 25.

Page 117
Half-life
It will take seven days for the radioactive output of the chemical to fall below 4 units. Your calculations should show the decrease as follows: 463 → 231.5 → 115.75 → 57.9 → 28.9 → 14.5 → 7.23 → 3.62.

Orang-utan puzzle
The number of orang-utans would fall below 2 500 after five years. Your calculations should show the following: 5 000 → 4 250 → 3 613 → 3 071 → 2 610 → 2 219.

Prize money puzzle
Method 2 is best as, although it starts off slowly, it produces almost twice as much money in the end. With method 1 you would get 651 notes after ten years and with method 2 you would get 1 133 notes.

Page 118
Square numbers
$37^2 = 1369$; $0.5^2 = 0.25$.

Square number puzzles
1. In each pair, one number is the reverse of the other and their squares are also the reverse of one another, for instance, $12^2 = 144$ and $21^2 = 441$. The same thing happens with several other numbers made from the digits 0, 1, 2 and 3. It will not work with numbers containing higher digits because the squares of numbers above 3 are larger than 10, so "carrying" upsets the pattern.
2. $1^2 = 1$; $11^2 = 121$; $111^2 = 12321$; $1111^2 = 1234321$. All these square numbers are palindromes. That means they are the same whether read backwards or forwards.
3. $5^2 = 25$
$15^2 = 225$
$25^2 = 625$
$35^2 = 1225$
$45^2 = 2025$
In this sequence the square numbers increase by 200, then 400, 600 and 800, and the next square number (55^2) will be 1 000 more, that is, 3 025.
4. $25^2 + 26^2 = 1301$
5. $3^2 + 6^2 + 7^2 = 2^2 + 3^2 + 9^2$
Each side of this equation totals 94, so it is true. All the other equations are true too.

Page 119
Ancient triangles
In the first triangle $9^2 + 12^2 = 225$ and $15^2 = 225$. The second triangle is also right-angled because $8^2 + 6^2 = 100$ and $10^2 = 100$.

Stone Age right-angles
The Stone Age measurements which give perfect right-angles are:
3, 4, 5 because $3^2 + 4^2 = 5^2$
8, 15, 17 because $8^2 + 15^2 = 17^2$
5, 12, 13 because $5^2 + 12^2 = 13^2$
12, 35, 37 because $12^2 + 35^2 = 37^2$
The others are very close indeed:
$41^2 + 71^2 = 6722$ and $82^2 = 6724$
$19^2 + 59^2 = 3842$ and $62^2 = 3844$
$8^2 + 9^2 = 145$ and $12^2 = 144$.

Pythagoras puzzles
In the first triangle the length of the third side is 1.5m because $1.5^2 + 3.6^2 = 3.9^2$. The second triangle has two of the same measurements, so its third side must be 3.9m.

Page 120
Opposites
1. The operations + and − are inverses of one another and so are × and ÷.
2. Here are the operations with their inverses: × 12 and ÷ 12; + 17 and − 17; − 7 and + 7; ÷ 10 and × 10.
3. The operation × 3 does reverse the effect of ÷ 3, but with some numbers, e.g. 14, the calculator cannot give an exact answer when dividing by 3 (14 ÷ 3 = 4.666 666 6), so when you multiply by 3 the result is not exactly 14, but it is very, very close (13.999 999 9).

Alternatives
Two inverses of ÷ 0.5 are × 0.5 and ÷ 2.

Inverses puzzle
These are the pairs of inverse operations: × 42 and ÷ 42; − 0.5 and + 0.5; + 42 and − 42; ÷ 5 × 8 and ÷ 8 × 5; × 1.6 and × 5 ÷ 8; + 1.6 and − 1.6; × 4.2 and ÷ 4.2.

The operations × 1 and − 0 are not really inverses because they both do the same thing (leave the number as it is). The operations ÷ 0 and × 0 have no inverses because it is not possible to divide by zero, and multiplying by zero reduces a number to nothing.

Page 121
Square roots
If 16 is the square root, the square number is 256.
If 16 is the square number, the square root is 4.
If 256 is the square number, 16 is the square root.

Square roots without a √ key
$\sqrt{70} = 8.367$

Page 122
Circular calculations
The Chinese value for π (3.141 592 9) is closest to the computer's figure. The other values are: Egyptian, 3.160 493 6; Babylonian, 3.125; Greek, 3.146 264 3; Archimedes, between 3.142 857 1 and 3.140 845; Indian, 3.141 6.

Using π
A can with a 16cm diameter would need a label at least 50.3cm long.

Earth puzzles
These are the answers using 3.142 for the value of π. If you used a different value your answers may vary slightly from these.
1. The circumference of the Earth is 39 700km to the nearest hundred km.
2. The length of the satellite's orbit is 40 343km. The calculation you need to do is π × 12 840, because the diameter of the orbit is: 100km + 12 640 + 100km = 12 480km.
3. The band around the satellite is 15.7m long.
4. If the band were mounted 1m from the surface of the satellite, the extra 2m on the diameter would make the length of the band 22m, so it would be 6.3m longer.
5. The extra length of rope needed to encircle the Equator at a height of 1m is also 6.3m. The answers to this and the previous question are the same because in each case you added 2m to the diameter, so the circumference became 2 × π longer.
6. The spaceship travels 13 440km in one orbit of the Moon (that is 8 × 1 680km). To work out how far its orbit is from the Moon you need to find the diameter of the orbit. The circumference of the orbit is 13 440km, so π × the diameter = 13 440 and the diameter is therefore 4 280 (to the nearest ten km). The diameter of the Moon is 3 480km so the distance of the spaceship from the Moon is ½(4 280 − 3 480), that is about 400km.

Division puzzle
Here are some five digit numbers starting with 7 which work in the same way as 38 125: 70 245; 72 605; 78 920; 76 520. There are several more.

An example of a six-digit number which works in the same way is 126 450 and a seven-digit number is 3 216 549. The only nine-digit number which works in this way is 381 654 729.

Page 124
Large number problems
2. If you split the subtraction as shown on the right, the top part of one of the sums is smaller than the bottom.

Examples
1. 32 707 793 946; 2. 244 148 291; 3. You need to split the sum into three parts. The answer is 2 184 354 064 327 273 546.

Page 125
Chessboard problem
The number of grains of rice on the last square of the chessboard would be approximately 9 200 000 000 000 000 000. That alone is more rice than the whole world can produce. Using the constant × 2 your calculations should show that on the eighth square there are 128 grains. The sixteenth square has 32 768 grains. After the 27th square, where there would be over 67 million grains of rice, the number exceeds the calculator display. To continue the calculation you need to cancel the overflow check, or convert the number to millions and re-enter it, (you will need to re-enter the constant too).

141

Page 126
Examples
1. 30; 2. 130; 3. 1; 4. 40.75; 5. 1.44; 6. 2.

$135 \div (15 - 6) + 5 = 20$
$135 \div 15 - (6 + 5) = -2$
$(135 \div 15) - 6 + 5 = 8$

Page 127
Brackets inside brackets
1. 1; 2. 10; 3. 0.6.

Reciprocals
The reciprocal of 0.5 is 2 and the reciprocal of 0.1 is 10.

The fractions $3/5$ and $1/3 + 1/5 + 1/15$ are equivalent because $3/5 = 0.6$ and
$$1/3 = 0.333\,333\,3$$
$$1/5 = 0.2$$
$$1/15 = \frac{0.066\,666\,6}{0.599\,999\,9}$$

Pyramid puzzle
The slope on the pyramid is 0.72. This value is the ratio, at any point on the slope, between the height and the horizontal distance from the corner. (A ratio is a way of expressing the relationship between two numbers by dividing them.)

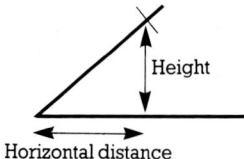

Egyptian builders made sure that the slopes on pyramids were uniform by calculating the ratio of $\frac{\text{horizontal distance}}{\text{height}}$, at lots of points on the slope and ensuring it was the same every time.

Page 128
Power puzzles
$7^2 = 49$; $2^{14} = 16\,384$; $2^{15} = 32\,768$;
$2^{30} = 1\,073\,741\,824$.

Problems with powers
The volume of the smaller box is 15.6 litres. The volume of the larger box is 125 litres, so it is eight times as big as the smaller one. Doubling the length of each side makes the volume eight times bigger because 8 is 2^3.

Which is bigger?
$2^5 = 32$ so it is bigger than $5^2 (= 25)$.
$4^3 = 64$ so it is smaller than $3^4 (= 81)$.

Leila's escape
The numbers Leila needs are 2 and 4 because $2^4 = 4^2$.

Page 129
Fractional powers
$9^{1/2}$ and $\sqrt{9}$ are both 3; $25^{1/2}$ and $\sqrt{25}$ are both 5; $64^{1/2}$ and $\sqrt{64}$ are both 8.

$60^{1/4} = 2.783\,157\,7$

Matching powers and roots
The complete table is

$64 = 2^6$	$81 = 3^4$	$256 = 4^4$	$9 = 9^1$
$\sqrt{64} = 2^3$	$\sqrt{81} = 3^2$	$\sqrt{256} = 4^2$	$\sqrt{9} = 9^{1/2}$

Powers in fractions
1. $1/5^3 = 0.008$; $1/6^3 = 0.004\,629\,6$; $1/9^2 = 0.012\,345\,7$; $1/10^4 = 0.000\,1$.
2. The numbers 5^{-3} and 0.2^3 are the same because 5^{-3} is the same as $1/5^3$ and $1/5 = 0.2$.

FIX key
$150 \div 7 = 21.43$ (to two decimal places); 21.429 (to three decimal places); 21.428 6 (to four decimal places).

$\$78 \div 11 = \$7$ and 9 cents.

Page 130
Converting numbers
1. 7.438×10^{11};
2. 9.23×10^9;
3. 8.02×10^{11};
4. 4.532×10^{10}.
5. 153 000 000 is 1.53×10^8, so it is bigger than 1.49×10^8.
6. 4.17^{10} is 41 700 000 000 which is bigger than 587 000 000.
7. 3.8×10^9 is about four times bigger than 9.3×10^8.
8. 1.9×10^6 is nearly 2 million and 9.1×10^5 is under 1 million.

Small number puzzle
Starting with the smallest, the order of size is: hydrogen atom, egg white protein molecule, 'flu virus, pneumonia bacteria, mumps virus, pin prick, paramecium protozoa.

Page 131
Puzzles
1. 9.75×10^{17}; 2. 1×10^{12}; 3. 200.
4. For a 70-year-old person the number of heartbeats would be approximately 2.58×10^9, that is 2 580 million.
5. A light year is 9.45×10^{12} kilometres (that is nearly ten million million).
6. The blood cell is 30 times larger than the bacteria cell.
7. The diameter of a 'flu virus is 250 times bigger than that of a hydrogen atom.
8. The flea could jump 70 metres.

SCI key
1. 1.50×10^1 or 15; 2. 5.26; 3. 1.18×10^8;
4. 3.32×10^{-3}.

Page 132
Statistics puzzles
The total age of all the 38 people in the group is 1 246 years. Their average age is 32.8 years.

Averages and statistics
For set 1, the sum of the numbers (Σ) is 594 and the average ($\bar{x}$) is 54. The standard deviation (σ) is 5.7. This is quite a small standard deviation because all the numbers are close to the average. They are all within the range 46-65.

For set 2, the sum of the numbers is 552 and the average is 55.2. The standard deviation is 42.1. This is a larger standard deviation because the numbers are spread out between 7 and 126.

Page 133
More statistics questions
1. The average age is over 30 because several of the people who are over 30 are very much older (e.g. 93, 66, 64) and these large numbers move the average up.
2. The percentage of people over 60 years old is 10.5%, and the percentage under 20 years old is 29%.
3. Judging by this sample 0.11 or 11% of TV programmes should be for children under ten.

Decimal maze
Here are two routes through the maze.

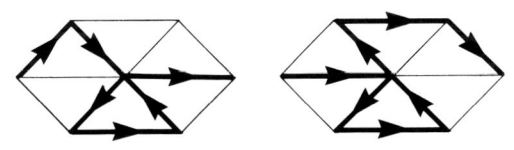

Page 134
Likely or unlikely
The probability of choosing a diamond from a pack of cards is 13/52 that is, 0.25 or 1 in 4.

Probability scale

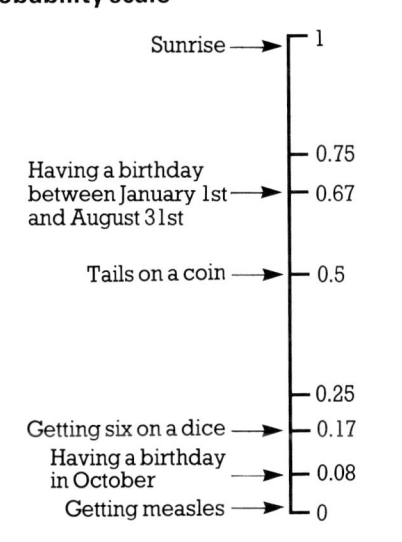

Sunrise → 1

Having a birthday between January 1st and August 31st → 0.75 / 0.67

Tails on a coin → 0.5

Getting six on a dice → 0.25 / 0.17

Having a birthday in October → 0.08

Getting measles → 0

Page 135
Aces puzzle
The probability of getting four aces is 3.69×10^{-6}, that is less than four in a million.

Factorials
6! is 720.

The calculations $4/52 \times 3/51 \times 2/50 \times 1/49$ and $\frac{4! \times 48!}{52!}$ are equal because: $4 \times 3 \times 2 \times 1$ is the same as 4! and $52 \times 51 \times 50 \times 49$ is the same as $\frac{52 \times 51 \times 50 \times 49 \times 48!}{48!}$ which is $52!/48!$.

Incidentally 52! is an enormous number. It is more than 8×10^{67}. This is a bigger number than there have been seconds since the world began.

Adding probabilities
The probability of the landing going wrong is 0.250 238 (that is, $1/5 + 1/35 + 1/60 + 1/200$), which is just outside the safety margin, (1 in 4 is 0.25).

Page 136
Mine rescue
The air-passage must be drilled at an angle of at least 32°. The calculation you need to do is as follows:
$80/130 = 0.615$ ← This is the tangent of angle A.
Pressing the inverse, then the tan key will give you the size of angle A (31.6°).

Trigonometry
The tangent of 29° would be slightly larger than 0.5317.

Sines and cosines
The sine of 28° is 0.469 and the cosine is 0.883. Your answers may not be exactly the same as these but they should be very close.

Page 137
Tower puzzle
The tower is 25.6 metres high. This is how to work it out:
$$\frac{\text{Height of tower}}{20} = \tan 52°,$$
so height of tower $= \tan 52° \times 20$.

Comet alarm
The comet will not collide with Earth. It will pass at a distance of 105km. This is how to work it out:
$$\sin 2° = \frac{d}{6\,400 + 180\,000}$$
so $d = \sin 2° \times 186\,400 = 6\,505$ (to the nearest km). The radius of the Earth is 6 400km so the comet's distance from the Earth will be $6\,505 - 6\,400 = 105$km.

Trig puzzles
In the first triangle side b is 4.2m and side c is 6.5m. In the second triangle angle x is 58° and angle y is 32°.

143

Index

accuracy
  on a calculator, 110
  three-figure, 62
acres, 15
acute angles, 20
addition, 25, 54, 62, 66, 78
  key, 53
algebra, 35
algebraic logic, 78-79
algorithms, 62
Al-Khowarismi, Mohamed, 35
All Clear key, 52
angles, 20-21, 80, 81
arc, 16, 36
Archimedes' value for π, 74
ARC key, 81
area, 14-15
  of circles, 16
arithmetic logic unit, 59, 61, 62
arithmetic means, 86-87
averages, 31-32, 86-87, 90, 84-85
axis, 32

Babylonians, 21
Babylonian value for π, 122
base 60 system, 41
base ten system, 40
base two system, 40
batteries, 53
bearings, 21
Bidder, George, 10
binary, 40-41, 60-61
bisecting (lines and angles), 37
brackets, 126-127
  keys, 78-79

calculations with several parts, 65, 78-79
calculators in space, 83
calculators, types of, 99
Calculator verses brain race, 62
calculator with a game, 79
calculator words, 63
change sign key, 52, 54, 67, 82, 85
Chinese value for π, 122
chord, 17, 36-37
circles, 16-17, 36-37, 74, 122
circuits, 60, 61, 76
circumference, 16, 74, 122
Clear Entry key, 100
compasses, 36
computer programs, 45
computers, 40-41, 83
computer's value for π, 122
cone, 17
congruent (shapes), 13
constant function, 66, 71, 73, 77
  for squaring numbers, 118
  how to use, 115
  key, 77
conversions, 66, 113
co-ordinates, 32, 34

cos (cosine), 80-81, 137
count-downs, how to do, 66
cube (power 3), 38
cube (solid), 18-19
cubing, 75, 82
cuboid, 19
cylinder, 17

Date of birth detector, 55
decimal code, 60
decimal fractions, 6-7, 25, 102, 106-109, 134
decimal point, 6, 7
decimal point key, 52
decimal system, 5, 6
decoder unit, 59, 61
degrees, 20, 21, 81
denominator, 24
Descartes, René, 32
Devi, Shakuntala, 11
diameter, 16, 37, 122
dice throwing, 42, 45
digits, 5, 40
display panel, 52, 53, 59, 64, 67, 74, 76, 83, 84
division, 7, 54, 65, 66, 78, 79, 106-107
  key, 5
  quick tips, 10-11
  using the memory for, 113
dodecahedron, 19
double function keys, 76

Earth, diameter of, 122, 123
Easter, how to calculate, 126
Egg white protein molecule, size of, 130
Egyptians, 4, 36, 127, 142
  Egyptian value for π, 122
electronic components, 60
ellipse, 17
encoder unit, 58, 61
entering a number, 54
enter key, 76
equals key, 52
equations, 34-35
equilateral triangle, 13
error alert, 67
Euclid, 36
Euler, Leonhard, 22
exponents, 84, 85, 131

faces, of solids, 18
factorials, 88, 89, 135
factors, 106, 138
financial calculator, 50
FIX key, 129
flag register, 58, 59, 61
'flu virus, size of, 130
formula,
  for area, 14
  for area of circle, 16
  for centigrade to Fahrenheit, 34
  for prime numbers, 39

for volume of cuboid, 19
fortune-telling calculator, 50
fractional powers, 129
fractions, 24-25, 57, 73, 106, 108-109, 134
  converting to percentages, 68
  powers in, 129
frequency graph, 30
function method (of solving equations), 35
functions,
  definition of, 52, 99
  inverse, 99, 120, 121

Gauss, Karl, 11
geometry, 36-37
gradients, 81
graphs, 30-35
Greek words, 12, 18, 36
Greeks, 17, 36
  Greek value for π, 122

hectares, 15
hexagon, 12
histogram, 30
hydrogen atom, size of, 130
Hypatia, 17
hyperbola, 17
hypotenuse, 72, 80

icosahedron, 19
Indian value for π, 122
indices (sing. index), 38
infinity, 5
instructions, in sums, 8
integrated circuit, 60
inverse,
  key, 76, 81, 82, 92, 99
  operations, 120-121
isosceles triangle, 13

keyboard, 58
  puzzles, 100-101

labels in sums, 8
large numbers on a calculator, 124-125, 130-131
LEDs (light emitting diodes), 53
liquid crystal, 53
logic, 3

mathematics, meaning of, 3
maths words, 48
matrix (matrices), 29
mean, 31
median, 31
memory, 51, 64-65, 66, 78, 86, 112-113
  keys, 52, 64, 65, 66, 77
  sums, 65
mental arithmetic, 10-11, 62
Middle Ages, 4, 8
minutes, 81
mistakes, how to spot, 55, 104,

144

138
mixed numbers, 57
mode, 31
modules, 51
motorway planning, 23
multiplication, 7, 54, 62, 66, 70, 71, 78, 82
    key, 53
    quick tips, 10-11
mumps virus, size of, 130

negative,
    exponents, 84, 85
    numbers, 8, 52, 54, 67
    powers, 35, 82
nested brackets, 79, 127
network, 22, 23
number of data key, 87
number path, 8-9
numbers, very large or very small, 84-85
number systems, 4-5, 40-41
number tricks, 63, 70, 71
numerator, 24

obtuse angles, 20
octahedron, 18
on/off switch, 53
overflows, 125

parabola, 17
parallel lines, 37
parallelogram, 13
    area of, 15
paramecium protozoa, size of, 130
pentagon, 12
percentages, 24-25, 51, 68-69, 90, 116-117, 133, 134
permanent memory, 58, 59, 61, 76
permutations, 88, 89
pi (π), 16, 122
    key, 52, 74-75
pie chart, 30
pin prick, size of, 130
plotting graphs, 32
pneumonia bacteria, size of, 130
polygons, 12
polyhedra (*sing.* polyhedron), 18-19
powers, 38-39, 40, 82-83, 128-129, 130, 131
    fractional, 129
    in fractions, 129
    negative, 129
predictions, 42, 91
prime numbers, 38-39
    computer programs, 45
probability, 42-45, 91, 134-135
probability scale, 43, 134
programmable calculators, 51, 83

protractor, 20, 37
pulses of electricity, 60
Pythagoras' theorem, 27, 72, 80, 119

quadrilateral, 12, 13
    area of, 15

radians, 81
radius, 16, 74
ratios, 26-27, 137
rearranging sums, 73, 78
reciprocal, 39, 72-73, 127
    key, 52, 72, 73, 82
recurring decimals, 107
rectangle, 13
    area of, 14
reflex angles, 20
regular polygons, 12
regular polyhedra, 18-19
remainders, 106, 139
Reverse Polish notation, 76
right angles, 20, 27, 119, 136-137
risks, how to calculate, 44, 83
Romans, 4-5
roots, 82, 121, 129
rough guesses, 55
rounding-off, 7, 56, 57, 110

sampling, 30
scalene triangle, 13
scanners, 58
scientific
    calculators, 99, 126, 130, 131, 132, 135, 137
    notation, 84-85, 130-131
SCI key, 131
second function, 76
seconds, 81
sector, 17
semi-circle, 17
sets, 28-29
shapes, naming, 12-13
significant figures, 7, 110, 131
silicon chip, 60-61
similar (shapes), 13
simple calculators, 99
simplifying fractions, 25
sin (sine), 80-81, 137
solar-powered calculator, 53
solids, 18
solving equations, 34-35
space shuttle, 83
sphere, 17, 75
square roots, 38, 62, 70-71, 72, 76
    key, 52, 70, 72, 121, 129
    without a square root key, 121
squares (power 2), 38, 70-71, 72, 76, 82, 118-119, 120
    key, 52, 70-71, 72
    without a square key, 71
squares (shapes), 13

standard deviation, 87, 90
    key, 86, 87, 132
standard form, 39, 130
statistics, 30-31, 86-91, 132-133
    keys, 86, 87
    register, 86
    register clear key, 86
Stonehenge, 119
subtraction, 25, 54, 65, 66, 78, 79
    key, 53
sum of data key, 87
sum of data squared key, 87
Sun, diameter of, 111

tally, how to do, 66
tangent,
    in trigonometry, 80-81, 136-137
    to circle, 17
tessellation, 13
tetrahedron, 18
theorem, meaning of, 20
three-dimensional (3-D) shapes, 18-19
topology, 22-23
transistors, 60
traversable, 22-23
tree diagrams, 43
triangle, 12, 13, 27, 72, 80-81, 119, 136-137
    area of, 15
trigonometry, 80-81, 136-137

units, of measurement, 15
user memory, 59, 64-65

vectors, 9
velocity, 9
Venn diagrams, 29
vertex (vertices), 22-23, 37
volume, 19, 27, 128

warning device, 67
watch (with calculator), 51

X register, 58, 59, 61, 62

Y register, 58, 59, 61, 62

zero, 8